INTRODUCTION TO RADIOCHEMISTRY

Introduction to Radiochemistry

David J. Malcolme-Lawes, BSc, PhD, CChem, FRIC

Department of Chemistry
Loughborough University of Technology
United Kingdom

A HALSTED PRESS BOOK

JOHN WILEY & SONS
New York

First published in Great Britain 1979 by
The Macmillan Press Ltd

Published in the U.S.A. by
Halsted Press, a Division of
John Wiley & Sons, Inc.
New York

Printed in Great Britain

Library of Congress Cataloging in Publication Data

Malcolme-Lawes, D J
Introduction to radiochemistry

"A Halsted Press book."
Includes indexes.
1. Radiochemistry. I. Title.
QD601.2.M33 541′.38 79–18096
ISBN 0 470–26783–6

Contents

Preface

Radiochemistry is the application of radioactivity to the solution of chemical problems. The importance of radiochemistry has increased dramatically over the last decade as radioactive materials have become more widely available, and as workers in many fields have taken advantage of the great sensitivity of radioactivity detection techniques. This book attempts to provide the reader with a basic understanding of radioactivity and of the most important techniques associated with the application of radioactivity to the chemical sciences. No previous knowledge of radiochemistry or nuclear physics is required, although it has been assumed that the reader will have a knowledge of chemistry to at least first-year undergraduate level or equivalent. It is hoped that the book will be found useful both by students in higher education, and by practising chemists, biochemists and biologists who may be able to employ radiochemical techniques in their work.

My thanks go to the many individuals and companies who provided information and demonstrated items of equipment during the writing of this book, and in particular to my colleagues at Loughborough University: Dr David Brown, who read the manuscript and made many suggestions for improvement; and Mr Graham Oldham, who created the interest in radiochemistry in this university and who is responsible for its excellent radiochemistry laboratory.

Loughborough, 1979 D.J.M.-L.

1 The Nucleus and Radioactivity

1.1 The nucleus

Atomic nuclei consist of positively charged protons and uncharged neutrons, particles which are collectively known as nucleons. Nucleons interact with one another through a short range attractive force, and it is this force which holds the nucleus together. Protons also interact with one another through a long range (inverse square law) repulsive force. Nuclear stability depends to a first approximation on the combination of the short range attractive forces and the longer range repulsive forces within the nucleus. As neutrons contribute mainly attractive forces to this sum, there is a sense in which the neutrons can be regarded as keeping the nucleus together. As the number of protons in the nucleus (the atomic number, Z) increases, so more neutrons are required to prevent the nucleus breaking apart under the strain of proton–proton repulsions.

Atoms are now known with atomic numbers up to about 108. For all of these elements a range of isotopes has been made or discovered, that is nuclei with a given number of protons but differing numbers of neutrons. The majority of these isotopes of the elements—or nuclides—are observed to be unstable with respect to other nuclides, and spontaneous transformations of these isotopes to more stable nuclides are seen. These spontaneous transformations are called radioactive decay, and a nuclide which undergoes radioactive decay is referred to as a radionuclide or a radioisotope.

Nuclides are most commonly referred to by the elemental symbol preceded by the mass number, A, for example, ^{12}C, ^{16}O, and ^{238}U. *Figure 1.1* shows many of the known nuclides arranged to show the number of protons and the number of neutrons in the nucleus. Those nuclides which are stable and do not undergo radioactive decay are shown in black squares, the others—the radionuclides— are shown in white squares. Clearly the stable nuclides lie in a relatively narrow band which extends from hydrogen, ^{1}H, at the bottom left of the diagram, to bismuth, ^{209}Bi, well up in the right hand portion of *Figure 1.1*.

The radionuclides are unstable, and this generally means that the total energy content of the nucleus is greater than that of the nearest stable nuclide. There are several routes which radionuclides may attempt to achieve stability, the actual route of radioactive decay for any particular radionuclide being dependent largely on the position of the nuclide with respect to the band of stable nuclides shown in *Figure 1.1.*

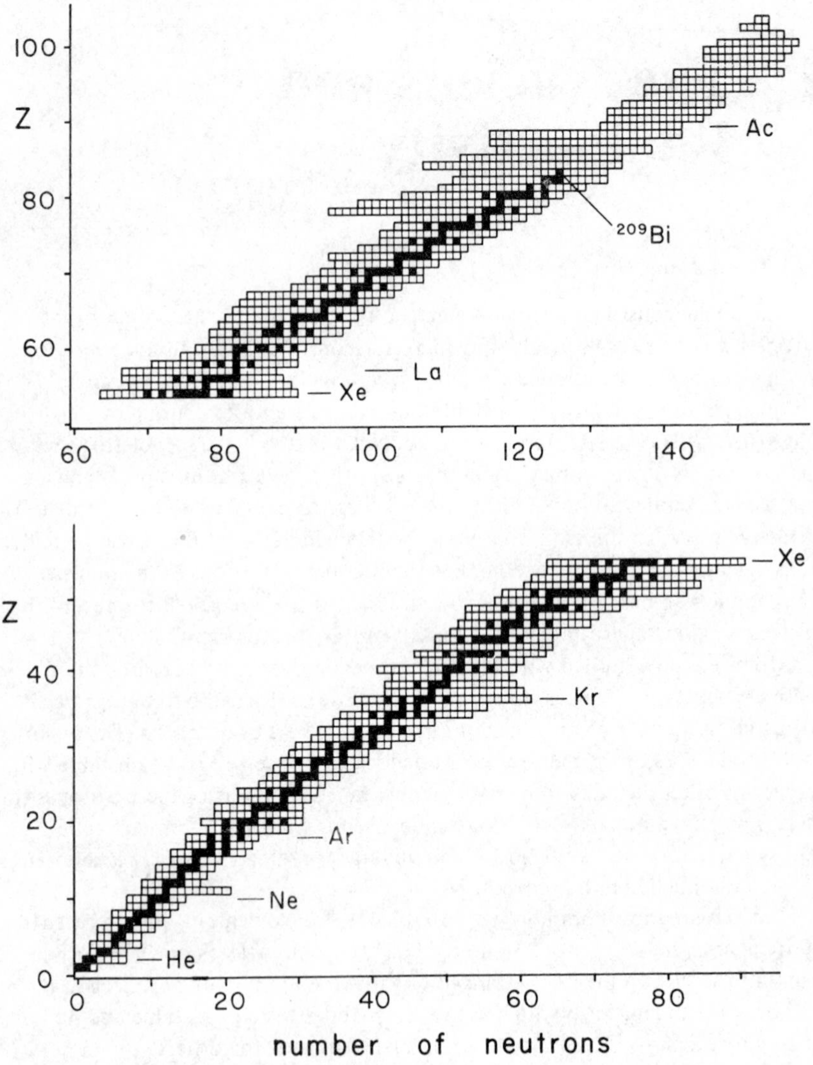

Figure 1.1 Most of the known nuclides are represented on this chart. Stable nuclides are shown as solid squares, ▪. The majority, shown as ▫, are unstable and undergo some form of radioactive decay.

1.2 β decay

The radionuclides of low mass number (say $A < 80$) almost invariably try to approach stability by one of several processes known collectively as β decay. The common feature of the β decay processes is that one nucleon within the nucleus is converted into the other type of nucleon. For example, a neutron, n, may be converted into a proton, p^+, a process known as β^- decay and which occurs when a nucleus spontaneously emits an electron, e^-. β^- decay may be written

$$n \rightarrow p^+ + e^-$$

(In fact an anti-neutrino is also emitted during β^- decay, but as these are extremely difficult to detect they are of no interest in radiochemistry. We will neglect the role of these particles in our discussion of β decay processes.)

A typical β^- decay is that of the widely used radionuclide ^{14}C, which can be written

$$^{14}C \rightarrow {}^{14}N + e^-$$

or, more concisely

$$^{14}C \xrightarrow{\beta^-} {}^{14}N$$

The other β decay processes result in the conversion of a proton into a neutron, either through the ejection from the nucleus of a positively charged electron (a positron), e^+, in which case the process is called β^+ decay,

$$p^+ \longrightarrow n + e^+$$

or through the capture by the nucleus of one of the atomic (extranuclear) electrons,

$$p^+ + e^- \longrightarrow n$$

in which case the process is referred to as electron capture (e.c.) (or K capture by some authors).

The net result of both β^+ decay and e.c. is that the number of protons in the nucleus decreases by one, and the number of neutrons increases by one. It is perhaps not surprising to find that many radionuclides that undergo one of these modes of decay also exhibit the other; in fact, the two modes usually compete with one another. An example of a radionuclide which undergoes both β^+ decay and e.c. is ^{18}F,

$$^{18}F \xrightarrow[(97\%)]{\beta^+ \text{ decay}} {}^{18}O$$

$$^{18}F \xrightarrow[(3\%)]{\text{e.c.}} {}^{18}O$$

The first point that should be noted about the three β decay processes is that the mass number of the nuclide remains unchanged on decay. The second point is that the β decay process results in the nuclide moving its position diagonally on the chart of the nuclides in *Figure 1.1*, by one square down and one to the right for β^+ decay and e.c., or by one square up and one to the left for β^- decay. As the driving force for radioactive decay is the desire of the nucleus to achieve stability, it should come as no surprise to find that nuclides which lie below and to the right of the band of stability in *Figure 1.1*, tend to undergo β^- decay. Such nuclides are frequently termed 'neutron rich', as they possess too many neutrons (or not enough protons) to be included in the band of stable nuclides. Nuclides which lie above and to the left of the band of stability–termed 'neutron deficient' nuclides–tend to approach stability by undergoing β^+ decay or e.c., and frequently by both processes.

For the external observer the consequence of a β decay process is that the radioactive atom emits a signal, which can be used to identify the type of β decay

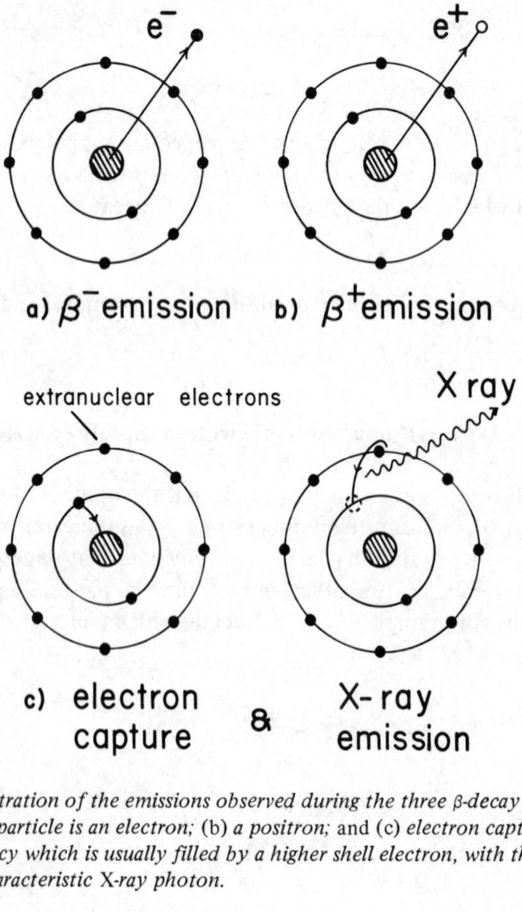

Figure 1.2 Illustration of the emissions observed during the three β-decay processes. (a) *The emitted particle is an electron;* (b) *a positron;* and (c) *electron capture leaves an inner shell vacancy which is usually filled by a higher shell electron, with the simultaneous emission of a characteristic* X-*ray photon.*

or to demonstrate the presence of the radioisotope. A nuclide undergoing β^- decay emits an electron (sometimes called a β^- particle), while β^+ decay results in positron emission. Electron capture does not directly involve the emission of any signal, although the capture by the nucleus of one of the innermost atomic electrons results in an inner shell vacancy–a situation usually relieved by an electronic transition from one of the higher electron shells with the simultaneous emission of an X-ray photon. The emissions observed during β decay are summarised schematically in *Figure 1.2*.

A β decay will allow an unstable nuclide to move one square nearer to the band of stability shown in *Figure 1.1*. For a nuclide several squares removed from stability a single β decay results in a product nuclide which is itself unstable (i.e. radioactive) and which in turn can undergo a second β decay. The sequence of one β decay following another can continue until a β decay results in the formation of a stable product nuclide. For example, the radionuclide ^{140}Xe requires a chain of four β^- decays before the stable nuclide ^{140}Ce is reached:

$$^{140}\mathrm{Xe} \xrightarrow{\beta-} {}^{140}\mathrm{Cs} \xrightarrow{\beta-} {}^{140}\mathrm{Ba} \xrightarrow{\beta-} {}^{140}\mathrm{La} \xrightarrow{\beta-} {}^{140}\mathrm{Ce}$$

In practice the most commonly available β^- decay isotopes are those which undergo a single β decay to form a stable product nucleus.

1.3 α decay

While the β^- decay processes are the most frequently observed routes through which radioactive nuclei approach stability, the higher mass number radionuclides, and particularly those that are heavier than the heaviest stable isotope (^{209}Bi), also use a different pathway. When a nucleus possesses too many nucleons for stability then no amount of β^- decay events will result in a stable product nucleus. The alternative is to eject some of the nucleons so that the number remaining in the nucleus is reduced. In fact this is achieved in most cases by the ejection of two protons and two neutrons in the form of the stable nucleus ^4He, otherwise known as an α particle. The form of radioactive decay in which an α particle is spontaneously ejected from a nucleus is called α decay.

Perhaps the most widely known example of α decay is that of ^{239}Pu (plutonium):

$$^{239}\mathrm{Pu} \rightarrow {}^{235}\mathrm{U} + \alpha$$

conventionally written as

$$^{239}\mathrm{Pu} \xrightarrow{\alpha} {}^{235}\mathrm{U}$$

Figure 1.1 demonstrates that α decay results in a move from one nuclide to a product nuclide two squares down the chart and two squares to the left. Some of the heavier radionuclides, which are not as massive as ^{209}Bi, apparently choose this decay pathway, in part because this can result in a more successful approach to stability than a single β decay.

1.4 Spontaneous fission

A third form of radioactive decay is occasionally observed among the heavier radionuclides. Unstable nuclei which contain too many nucleons can attempt to relieve this instability by splitting into two smaller nuclei. This process is known as spontaneous fission (s.f.) and results in the parent nucleus splitting into two roughly equal mass parts—forming two product nuclei (called fission products) which are themselves usually radioactive. Spontaneous fission is often accompanied by the ejection of one or two neutrons from the fissioning nucleus.

When radionuclides decay by spontaneous fission the fission products are found to include dozens of nuclides in the mass range 80–160, so an example of spontaneous fission will be only one of a large number of possible fissions that a given parent radionuclide can undergo. One example is the spontaneous fission of ^{252}Cf (californium), an isotope which is becoming increasingly popular as a source of the neutrons emitted during fission.

$$^{252}Cf \rightarrow {}^{120}Cd + {}^{130}Sn + 2n$$

1.5 γ emission

The modes of radioactive decay considered so far have involved a radionuclide attempting to relieve its instability by rearranging or reducing its complement of nucleons. These processes all result in a product or daughter nuclide which is an isotope of an element chemically different from the parent radionuclide. In fact nuclei can also exist in excited states which are to some extent analogous to the electronically excited states in which atoms and molecules can exist. For atoms and molecules, excited states are produced during chemical reactions or by the

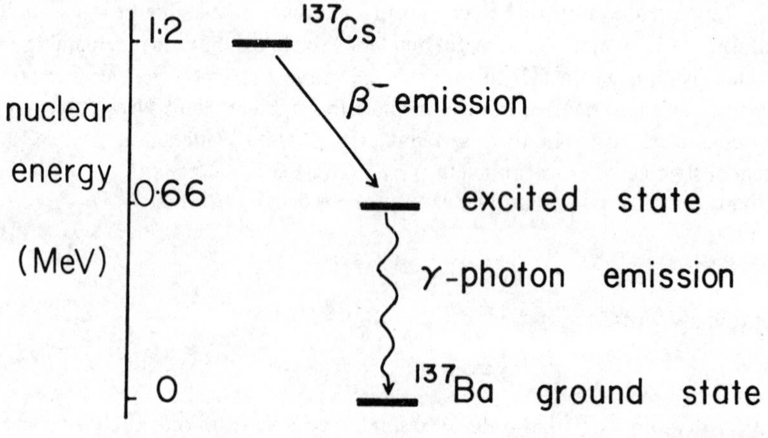

Figure 1.3 Simplified decay scheme for ^{137}Cs *showing the production of an excited state of* ^{137}Ba *by the* β^- *decay of* ^{137}Cs*. The excited state relaxes to the ground state of* ^{137}Ba *by emitting a 0.66 MeV* γ *photon. Nuclear energy is measured relative to the* ^{137}Ba *ground state in this example.*

excitation of the stable ground state following the absorption of electromagnetic radiation (i.e. photons). These excited states may then 'relax' to the ground state by emitting photons. Analogous processes may be used to generate excited state nuclei, although the most readily available sources of these excited states are the α and β radioactive decay processes. The majority of α and β decays result in excited state daughter nuclei, and these relax to the ground state for the daughter nucleus by the emission of one or more photons. This ground state may be stable or radioactive.

Photons emitted during the relaxation of excited state nuclei are called γ photons. γ radiation is usually of very short wavelength ($<$ 1 nm), although the distinction between γ rays and X rays is based on the photon source rather than the wavelength. A typical γ emission process is that which occurs immediately following the β^- decay of the isotope ^{137}Cs to an excited state of ^{137}Ba. The relevant energy levels of the nuclei involved are shown in *Figure 1.3.* In this example virtually all of the β decays result in the same excited state of the daughter nucleus, so that when relaxation to the ground state occurs the γ photons emitted all have the same energy (or wavelength). However, in many radioactive decays several different excited states of daughter nuclei are produced, so that γ photons of several different energies are emitted. The subject of γ emissions will be considered in a later chapter.

1.6 The decay law

In our discussion of radioactive decay we have not so far considered the time dimension. In fact, if we could isolate a single unstable atom we could not predict when it would undergo radioactive decay. However, when one collects a very large number of atoms, N_o, of a particular radioisotope, then after a time, t, the number remaining, N_t, is found to be

$$N_t = N_o \exp(-\lambda t) \tag{1.1}$$

where λ is a constant and is a characteristic of the radionuclide. Equation 1.1 is known as the radioactive decay law, and the constant λ is known as the decay constant of the radionuclide.

In view of the involved physical interpretation of the decay constant most radiochemists prefer to think about rates of decay in terms of the time required for half of a collection of radioactive atoms to decay, that is for

$$N_t = \tfrac{1}{2} N_o$$

From the decay law it follows that this occurs after a time $t_{1/2}$, which may be obtained from

$$\tfrac{1}{2} N_o = N_o \exp(-\lambda t_{1/2})$$

or

$$t_{1/2} = \frac{-\ell n 1/2}{\lambda} = \frac{0.693}{\lambda}$$

The time $t_{1/2}$ is known as the half-life of the radionuclide, and it is this characteristic property of the nuclide which is tabulated—the closely related decay constant being evaluated only as required.

The half-life of a radionuclide is related to the instability of the nucleus—the greater the instability the shorter is the half-life. For this reason the half-lives of the radionuclides tend to decrease as we move away from the band of stable nuclei on the chart of the nuclides (*Figure 1.1*). Regrettably the distance from the band of stability can only be taken as a very rough guide to half-life, because in practice the stability of a nucleus is a very complex function of the number of nucleons it contains. The half-lives of the radionuclides range from fractions of seconds to hundreds of millions of years.

The majority of the radionuclides decay either by α decay or by β decay, and so most have a single characteristic half-life. However, some of the heavier nuclides can decay both by the α decay and by spontaneous fission, and in such cases there may be a different half-life associated with each mode of decay. For example, the nuclide ^{252}Cf decays as follows:

$$^{252}Cf \xrightarrow[t_{1/2} = 2.73 \text{ yr}]{\alpha \text{ decay}} {}^{248}Cm$$

$$^{252}Cf \xrightarrow[t_{1/2} = 85.5 \text{ yr}]{\text{s.f.}} \text{fission products.}$$

Of course, a quantity of ^{252}Cf still disappears with an effective half-life of slightly less than 2.73 years, 2.65 years in fact.

Excited state nuclei also have a characteristic half-life associated with their relaxation and consequent γ emission. In general these are relatively short periods of time, ranging from a few hours down to below the picosecond (10^{-12} s) region, at which point measurement becomes difficult and the result irrelevant for our purposes. In many applications, where excited state nuclei with $t_{1/2} < 10^{-9}$ s are produced by α or β decay of a parent nuclide, the γ emission can be regarded as simultaneous with the α or β decay process. Thus the β^- decay of ^{24}Na is followed so rapidly by a γ emission from the excited state ^{24}Mg daughter that the two processes are essentially simultaneous, and ^{24}Na is often described as a β^- and γ emitting isotope.

Other excited state nuclei have half-lives which are sufficiently long to allow relatively simple distinction between the γ emission and the α or β decay process which produced the excited state. Such excited states are normally described as metastable states, and the excited state nuclide distinguished from the ground state by the addition of the letter m to the mass number. For example, the metastable state of the radionuclide ^{69}Zn is written ^{69m}Zn. In this case the metastable state has a characteristic half-life for γ emission, and the ground state a different half-life for its own β^- decay, such as

$$^{69m}\text{Zn} \xrightarrow[t_{1/2}\,=\,13.9\text{ h}]{\gamma} {}^{69}\text{Zn} \xrightarrow[t_{1/2}\,=\,55\text{ min}]{\beta^-} {}^{69}\text{Ga}$$

The half-life of a metastable state may be longer than that of the ground state nuclide, as in the ^{69}Zn case, or shorter, as for the widely used radionuclide ^{99m}Tc (technetium)

$$^{99m}\text{Tc} \xrightarrow[t_{1/2}\,=\,6\text{ h}]{\gamma} {}^{99}\text{Tc} \xrightarrow[t_{1/2}\,=\,2.1\times10^5\text{ yr}]{\beta^-} {}^{99}\text{Ru}$$

Alternatively the ground state may be stable, as it is, for example, in the case of ^{137}Ba:

$$^{137}\text{Cs} \xrightarrow{\beta^-} {}^{137m}\text{Ba} \xrightarrow[t_{1/2}\,=\,2.6\text{ min}]{} {}^{137}\text{Ba (stable)}$$

Metastable states are also called isomeric states of the nuclide, and the relaxation of a metastable state referred to as an isomeric transition (I.T.).

1.7 Activity

For a given sample of radioactive material an important physical property is the number of decays per second occurring within the sample. This quantity is called the activity of the sample and is recorded in units of Becquerels, Bq, (named after the discoverer of radioactivity) where 1 Bq = 1 radioactive disintegration per second (older literature may use the unit of d.p.s.) For example, 1 mg of pure ^{14}C contains

$$\frac{10^{-3} \times L}{14} = 4.302 \times 10^{19} \text{ atoms of } ^{14}\text{C}$$

(where *I* is the Avagadro Number).

The half-life of ^{14}C is 5,730 years, so the decay constant

$$\lambda_{14\text{C}} = \frac{0.693}{5730}\ \text{yr}^{-1}$$

$$= 3.835 \times 10^{-12}\ \text{s}^{-1}$$

Differentiating the decay law, equation (1.1) we obtain

$$\frac{dN}{dt} = -\lambda_{14\text{C}}\, N_0 \exp(-\lambda_{14\text{C}}\, t) = -\lambda_{14\text{C}}\, N$$

which is the rate of change of the number of ^{14}C atoms with time, the negative sign indicating a reduction in the number of atoms.

With N set equal to the number of ^{14}C atoms in our 1 mg sample, the rate of decay is

$$3.835 \times 10^{-12} \times 4.302 \times 10^{19}\ s^{-1}$$

$$= 2.151 \times 10^{8}\ s^{-1}$$

Thus the activity of 1 mg of ^{14}C is 2.151×10^8 Bq.

Before 1980 virtually all radioactivities were described in units of Curies—after the discoverers of radium. The Curie has the symbol Ci and

$$1\ \text{Ci} = 3.7 \times 10^{10}\ \text{Bq}$$

Hence our 1 mg ^{14}C sample had an activity of

$$\frac{2.151 \times 10^{8}}{3.7 \times 10^{10}} = 5.81 \times 10^{-3}\ \text{Ci or } 5.81\ \text{mCi}$$

Clearly there is a very great difference in magnitude between values of activity expressed in the Curie and in the modern Becquerel.

Not surprisingly there is a close relationship between the half-life of a radionuclide and the mass of that material which has an activity of, say, 1kBq—a fairly common level of activity used for laboratory studies. From the calculation above it should be clear that the mass of material which has an activity of 1 kBq, m_1, is given by

$$m_1 = \frac{10^3\ M}{L} \times \frac{t_{1/2}}{0.693}\ \text{g}$$

where M is the Molar mass of the material containing the radionuclide and $t_{1/2}$ is in seconds. Values of m_1 calculated for several radionuclides with a range of half-lives are given in *Table 1.1*. The masses give some indication of how small are the quantities routinely handled in the radioisotope laboratory.

Table 1.1 Mass of 1 kBq of some radionuclides

Radionuclide	$t_{1/2}$	*Mass of* 1 kBq (g)
^{238}U	4.5×10^9 yr	8.09×10^{-2}
^{14}C	5730 yr	6.06×10^{-9}
^{3}H	12.5 yr	2.83×10^{-12}
^{125}I	60 d	1.55×10^{-12}
^{18}F	1.9 h	2.95×10^{-16}
^{15}O	2.1 min	4.54×10^{-18}

1.8 Carriers

The handling of very small quantities of material can generate a number of problems. For example sub-picogramme quantities of a compound can easily become adsorbed on laboratory glassware or on solid phase material such as precipitates or filter papers. For this reason radioactive compounds are often mixed with quantities of non-radioactive but chemically identical material known as carrier. The use of carrier ensures that only a very small fraction of molecules of the compound contain the radioisotope, so that if small amounts of the compound are lost through adsorption and so on, the amount of radioactivity lost will be negligible.

The masses of 1 kBq of the elements in *Table 1.1* were calculated on the assumption that no carrier had been added to the radioisotopes. Those materials were carrier-free. When carrier is added to a radioactive substance a measure of the radioactivity per unit mass of the total material is known as the specific activity. Specific activity is usually quoted in $Bq\,g^{-1}$ or $Bq\,mol^{-1}$ for compounds, although $Bq\,dm^{-3}$ is also used for solutions. Other units which appear in pre-1980 literature are $Ci\,g^{-1}$, $Ci\,mol^{-1}$ and $Ci\,dm^{-3}$.

A radioisotopically labelled compound which is carrier-free offers the highest attainable specific activity for the compound labelled with the specified radioisotope, although this maximum specific activity will vary with different radioisotopes. For example CH_3I can be obtained labelled with ^{14}C, ^{3}H or ^{125}I. Carrier-free ^{14}C labelled CH_3I–written as $^{14}CH_3I$ or ^{14}C-iodomethane–has a specific activity 2.3×10^{12} $Bq\,mol^{-1}$, whereas carrier free $CH_2{}^{3}HI$ has a specific activity of 1.08×10^{15} $Bq\,mol^{-1}$, and carrier free $CH_3{}^{125}I$ reaches 8.06×10^{16} $Bq\,mol^{-1}$.

Bibliography

Friedlander, G., Kennedy, J. W. and Miller, J. M., *Nuclear and Radiochemistry*, J. Wiley & Sons, New York (1964)

McKay, H. A. C., *Principles of Radiochemistry*, Butterworths, London (1971)

Haissinsky, M., *Nuclear Chemistry and its Applications,* Addison-Wesley, Reading MA (1964)

Wilson, B. J., *The Radiochemical Manual,* The Radiochemical Centre, Amersham (1966)

Lederer, C. M., Hollander, J. M. and Perlman, I., *Table of Isotopes,* J. Wiley & Son, New York (1967)

2 Sources of Radioisotopes

2.1 Naturally occurring radioisotopes

The Universe is only about 1.3×10^{10} years old, so radionuclides which were created along with the stable nuclides and which have a half life of 10^8 years or more are still to be found. The radioisotopes which fall into this category are collected in *Table 2.1*, along with approximate values of their half-lives. Some of these radionuclides occur together with their stable isotopes in many parts of the natural world, including (in the case of ^{40}K) our own bodies. Others, for which there are no corresponding stable isotopes, occur in mineral deposits or seawater, and provided man with his first examples of radioactivity.

Table 2.1 Naturally occurring very long lived radionuclides

Isotope	*Half-life (yr)*
^{40}K	1.27×10^9
^{147}Sm	1.2×10^{11}
^{176}Lu	2.2×10^{10}
^{232}Th	1.39×10^{10}
^{238}U	4.5×10^9
^{235}U	7.1×10^8

Several of the naturally occurring very long lived radionuclides decay to shorter lived radioactive daughters. These in turn may decay to produce other radioisotopes, and the sequence may continue and give rise to a series of relatively short lived radionuclides before ending with a decay to a stable isotope.

One of these naturally occurring decay series is shown in *Figure 2.1*. This series begins with the very long lived ^{238}U and produces a large number of radioactive daughters before terminating in isotopes of lead. As a result of this series of radioactive decays natural occurrences of ^{238}U—such as the minerals pitchblende and uranite (both essentially UO_2)—contain quantities of all the other nuclides in the decay series. Because the decay processes have been occurring for a time period which is long compared with the half-life of all the members of the series except the parent ^{238}U and the stable lead isotopes, the shorter lived members of the series may be regarded as being in dynamic equilibrium with

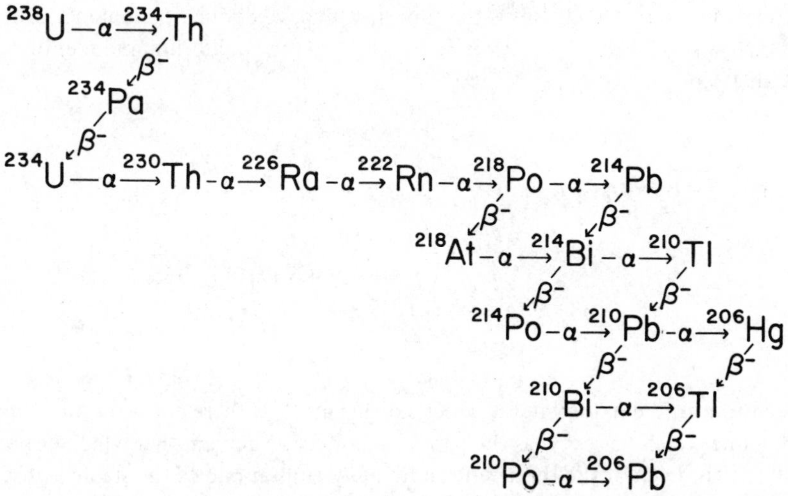

Figure 2.1 The uranium decay series

one another, and the relative concentrations of each radionuclide may be calculated using the decay law. For example, the first two decays in the ^{238}U series are

$$^{238}\text{U} \xrightarrow[t_{1/2}\ =\ 4.5\ \times\ 10^9\ \text{yr}]{\alpha} {}^{234}\text{Th} \xrightarrow[t_{1/2}\ =\ 24.1\ \text{d}]{\beta} {}^{234\text{m}}\text{Pa}$$

If we take the number of ^{238}U atoms currently present in a natural uranium sample as N_0, then the number of ^{234}Th atoms formed after time t, N^+_{Th}, is given by

$$N^+_{\text{Th}} = N_0\ (1 - \exp\ (-\ \lambda_{\text{U}} t))$$

where λ_{U} is the decay constant of ^{238}U.

The rate of formation of ^{234}Th is thus,

$$\frac{\text{d}N^+_{\text{Th}}}{\text{d}t} = N_0 \lambda_{\text{U}} \exp\ (-\lambda_{\text{U}} t)$$

The decay law applied to ^{234}Th allows calculation of the number of ^{234}Th decays after time t, N^-_{Th}, as

$$N^-_{\text{Th}} = N^{\text{eq}}_{\text{Th}}\ (1 - \exp\ (-\ \lambda_{\text{Th}} t))$$

where $N^{\text{eq}}_{\text{Th}}$ is the equilibrium number of ^{234}Th atoms in the sample, and λ_{Th} is the decay constant for ^{234}Th.

Thus the rate of decay of ^{234}Th is

$$\frac{\text{d}N^-_{\text{Th}}}{\text{d}t} = N^{\text{eq}}_{\text{Th}}\ \lambda_{\text{Th}} \exp\ (-\ \lambda_{\text{Th}} t)$$

At equilibrium the rate of formation of ^{234}Th will be equal to its rate of decay, thus

$$N^{\text{eq}}_{\text{Th}}\ \lambda_{\text{Th}} \exp\ (-\ \lambda_{\text{Th}} t) = N_0 \lambda_{\text{U}} \exp\ (-\ \lambda_{\text{U}} t)$$

We may now set $t = 0$, as this is the time at which we defined the number of ^{238}U atoms to be N_0. This allows us to calculate the equilibrium number of ^{234}Th atoms:

$$N_{Th}^{eq} = N_0 \frac{\lambda_u}{\lambda_{Th}}$$

or, since $\lambda = 0.693/t_{1/2}$

$$N_{Th}^{eq} = N_0 \frac{t_{1/2}\,(\mathrm{Th})}{t_{1/2}\,(\mathrm{U})} = N_0 \frac{24.1}{4.5 \times 10^9 \times 365}$$

that is

$$N_{Th}^{eq} = 1.47 \times 10^{-11} \times N_0$$

In addition to the natural decay series starting at ^{238}U there are two other major series, one which has ^{235}U as the parent radionuclide and another which begins with ^{232}Th. Each of the three series ultimately ends at one of the stable isotopes of lead, and full details of the series may be found in the standard texts on nuclear chemistry, some of which are mentioned in the bibliography.

2.2 Nuclear reactions

We have seen how some of the naturally occurring radionuclides have been created by the spontaneous decay of very long lived parents, which themselves are as old as most of the stable isotopes in the Universe. However, there are other naturally occurring radionuclides on Earth and some of these have come into being as a result of nuclear reactions.

A nuclear reaction is a reaction between a nucleus and a nucleon (or another nucleus) which results in the formation of a new nucleus. There are several natural sources of nucleons which can cause nuclear reactions on or near the Earth. One such source is the spontaneous fission of very long lived radionuclides such as ^{238}U, which, in addition to producing the radioactive fission products found in small quantities along with natural occurrences of uranium, gives rise to two or three neutrons per fission. These 'fission neutrons' can then react with other nuclei to produce new nuclides. An example which is of particular significance to man is the reaction between neutrons and ^{238}U nuclei:

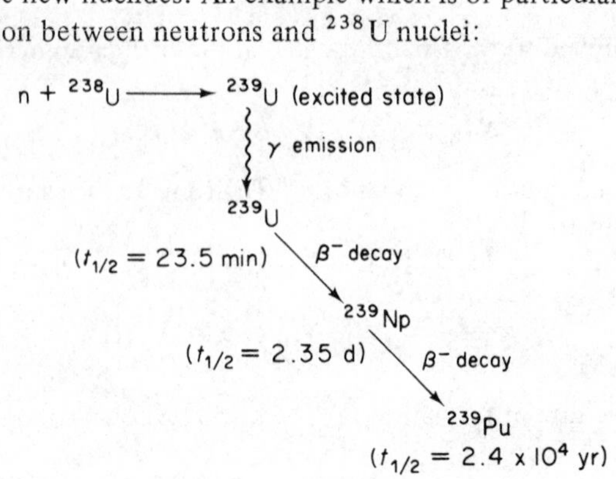

In this case the nuclear reaction between a neutron and ^{238}U nucleus results in a very short lived excited state ^{239}U nucleus, which decays to its ground state by emitting a γ photon. This particular sequence of events, that is, neutron absorption and γ emission, is very common and is usually written in the more concise shorthand of nuclear reactions as

$$^{238}U\ (n, \gamma)\ ^{239}U$$

and the reaction called an (n, γ) reaction. The same kind of shorthand is used for most other classes of nuclear reaction. The reactant nucleon and the emitted nucleon or photon are written between brackets and separated by a comma. The symbol of the reactant nuclide is written before the brackets and that of the product nuclide directly after the brackets.

The majority of stable nuclei undergo (n, γ) reactions, and some of the naturally occurring radionuclides have been created in this manner. However, there is a second important source of nucleons in our environment and which leads through nuclear reaction to the generation of radioisotopes. Cosmic rays are showers of fast moving nucleons, electrons and other fundamental particles. The atmosphere of our planet is continually bombarded by these particles and nuclear reactions are continually occurring between these nuclear particles and the atomic nuclei of atmospheric constituents. An example of this source of natural radioactivity results from the reaction between neutrons and nitrogen nuclei, ^{14}N, which produces the radionuclide ^{14}C. This reaction may be written

$$^{14}N\ (n, p)\ ^{14}C$$

and is clearly accompanied by the emission of a proton. ^{14}C produced in this manner enters the planet's carbon cycle from which some carbon may be trapped in a form which ceases to exchange with atmospheric carbon—in man made artefacts, and so on. As ^{14}C has a half-life of 5,730 years, the determination of the ^{14}C to (stable) ^{12}C ratio has become a useful technique for estimating the age of objects on the 1000–10 000 year time scale, and is known as radiocarbon dating. The $^{14}C/^{12}C$ ratio in the atmosphere is approximately 1.2×10^{-12}, which corresponds to an activity from ^{14}C of ~0.23 Bq per gramme of atmospheric carbon.

While the natural sources of radioisotopes are of general interest and of considerable importance in many fields, the vast majority of radioisotopes used in the laboratory are man-made. Almost without exception these are produced by nuclear reactions using man-made sources of nucleons.

2.3 Man-made radionuclides

The first radionuclide to be artificially produced was ^{30}P, obtained by Pierre and Marie Curie in 1934 using the nuclear reaction

$$^{27}Al\ (\alpha, n)\ ^{30}P$$

In this reaction α particles (from the α decay of ^{210}Po) react with stable ^{27}Al, a neutron is emitted and the short lived ($t_{1/2} \sim 2.6$ min) ^{30}P product nuclide is formed. For the positively charged α particle to enter the positively charged aluminium nucleus a very high Coulombic repulsion barrier must be overcome, and a very energetic α particle is required. The α particles emitted during α decay are usually quite energetic—again partly because of Coulombic repulsion between the nucleus and the departing α particle—and those from ^{210}Po are adequate in this case. However, even in 1934 high-voltage ion accelerators had been devised which could produce fairly high energy charged particles, and it was the subsequent development of these machines that allowed radioisotopes of most of the elements to be obtained in useful quantities.

Most of the man-made radionuclides created by charged particle nuclear reactions are produced using protons, deuterons (deuterium nuclei) or α particles (^4He nuclei) which have been accelerated to very high energies in a cyclotron. A cyclotron is essentially a high-voltage ion accelerator in which the ions are constrained to follow a spiral pathway by the presence of a magnetic field applied perpendicular to the plane of ion motion. (The principles of operation of the cyclotron and related accelerators are described in the texts given in the bibliography). Some examples of radionuclides commonly produced at cyclotron installations are given in *Table 2.2* along with the nuclear reactions used in their production.

Table 2.2 Typical cyclotron produced radionuclides

Radionuclide	*Half-life*	*Production reaction*	*Decay mode*
^{11}C	20.3 min	$^{14}N\ (p, \alpha)\ ^{11}C$	β^+
^{18}F	110 min	$^{20}Ne\ (d, \alpha)\ ^{18}F$	β^+
^{22}Na	2.58 yr	$^{24}Mg\ (d, \alpha)\ ^{22}Na$	β^+ and γ
^{57}Co	270 d	$^{58}Ni\ (p, 2p)\ ^{57}Co$	e.c., β^+ and γ
^{67}Ga	78 h	$^{67}Zn\ (p, n)\ ^{67}Ga$	e.c. and γ
^{111}In	2.8 d	$^{111}Cd\ (p, n)\ ^{111}In$	e.c. and γ

In general the radionuclides generated by charged particle bombardment are neutron deficient isotopes, and so tend to be those isotopes which decay by β^+ emission or e.c. Conversely most of the β^+ emitting or e.c. isotopes available commercially are produced by cyclotron bombardment. However, a cyclotron is an expensive machine and has no other major function than the initiation of nuclear reactions, hence the β^+ emitting and e.c. isotopes tend to be fairly expensive—a non-trivial consideration when planning radioisotope work.

A second major source of man-made radionuclides has become available since the development of nuclear fission reactors during the 1950s. A nuclear fission reactor is a structure in which a nuclear 'fuel', such as ^{235}U, undergoes fission and produces heat, fission products and fission neutrons. Thus a nuclear reactor can be used as a plentiful source of neutrons, and through (n, γ) and (n, p) reactions a very wide range of radioisotopes can be produced. Some of the most widely used reactor produced isotopes are given in *Table 2.3*, and it should be noted that these include some fission product isotopes recovered from the used reactor fuel itself.

In general, reactor produced radionuclides are neutron rich isotopes which tend

Table 2.3 Typical reactor produced radionuclides

Radionuclide	*Half-life*	*Production reaction*	*Decay mode*
^{3}H	12.5 yr	$^{6}Li\ (n, \alpha)\ ^{3}H$	β^-
^{14}C	5730 yr	$^{14}N\ (n, p)\ ^{14}C$	β^-
^{32}P	14 d	$^{32}S\ (n, p)\ ^{32}P$	β^-
^{60}Co	5.3 yr	$^{59}Co\ (n, \gamma)\ ^{60}Co$	β^- and γ
^{90}Sr	28 yr	Fission product	β^-
^{131}I	8 d	$^{130}Te\ (n, \gamma)\ ^{131}Te \xrightarrow{\beta^-} {}^{131}I$	β^- and γ
^{137}Cs	30 yr	Fission product	β^-

to decay by β^- emission. Conversely most of the commercially available β^- emitters have been produced by reactor irradiation or as fission products. Nuclear reactors (as distinct from nuclear power stations) are relatively inexpensive and frequently have functions other than isotope production. Consequently β^- emitting radionuclides tend to be cheaper than the e.c. and β^+ emitters.

The two techniques outlined above are the most widely used routes to the man-made radionuclides, and nearly all commercially available radioisotopes will have been prepared in one of these ways. However, there are several alternative routes which are of some importance, particularly for the production of relatively small quantities of short-lived isotopes at laboratories which do not have the facilities of a neighbouring reactor or cyclotron. Two of these will be mentioned because of the relatively low costs associated with these methods of radionuclide production.

2.4 Neutron generators

The first method involves a source of high energy (or 'fast') neutrons. Relatively inexpensive accelerators are available for the acceleration of deuterium ions sufficiently to enable the fusion reaction

$$^{3}H\ (d, n)\ ^{4}He$$

to occur when the deutron beam strikes a solid target containing tritium (^{3}H). This reaction produces neutrons which have a high kinetic energy of about 14 MeV (14×10^6 electron volts). Purpose built accelerators for use in the production of fast neutrons are called neutron generators and are available from a number of companies including Multivolt Ltd, Philips and Kaman Nuclear Inc.

The fast neutrons emitted by neutron generators can induce several nuclear reactions which do not occur with the much slower neutrons available at most fission reactors. Typical of the fast neutron (n*) reactions is the (n*, 2n) reaction in which one high energy neutron entering the reactant nucleus cause the ejection of two neutrons, leaving a product nucleus which is neutron deficient. Thus the radionuclides most readily produced by fast neutron irradiation tend to decay by e.c. or β^+ emission. Examples are

$$^{35}Cl\,(n^*, 2n)\,^{34m}Cl \qquad (t_{1/2} \sim 30\ \text{min})$$

$$^{19}F\,(n^*, 2n)\,^{18}F \qquad (t_{1/2} \sim 110\ \text{min})$$

$$^{14}N\,(n^*, 2n)\,^{13}N \qquad (t_{1/2} \sim 10\ \text{min})$$

Alternatively the fast neutrons can be 'moderated' (slowed down) to produce slow neutrons which can then be used, in much the same way as reactor neutrons, to induce (n, γ) and (n, p) type reactions leading to the production of neutron rich radionuclides. In general, neutron generators provide less neutrons per second than fission reactors, so the quantity of radioisotopes obtainable with these accelerators is small, and their main application in this area is for the production of short-lived isotopes.

2.5 Nuclide generators

The second means of obtaining a limited number of short-lived isotopes in a laboratory some distance from a reactor or cyclotron uses a nuclide generator. A nuclide generator is little more than a supply of a moderately long-lived parent radioisotope which decays to produce a required short-lived isotope. Examples include generators based on the following decay processes:

$$\underset{(t_{1/2} = 67\ \text{h})}{^{99}Mo} \xrightarrow{\beta^-} \underset{(t_{1/2} = 6\ \text{h})}{^{99m}Tc}$$

$$\underset{(t_{1/2} = 4.7\ \text{h})}{^{81}Rb} \xrightarrow{\text{e.c. and } \beta^+} \underset{(t_{1/2} = 13\ \text{s})}{^{81m}Kr}$$

$$\underset{(t_{1/2} = 118\ \text{d})}{^{113}Sn} \xrightarrow{\text{e.c.}} \underset{(t_{1/2} = 1.7\ \text{h})}{^{113m}In}$$

In most nuclide generators the parent isotope is adsorbed on a support material, such as an ion exchange resin, which is packed into a short tube. The tube is normally kept filled with an aqueous buffer solution in which the short-lived daughter nuclide dissolves as it forms, leaving the parent nuclide adsorbed on the support material. The daughter nuclide may be obtained simply by pumping the aqueous buffer out of the generator tube.

2.6 Availability of radioisotopes

The majority of users of radioisotopes do not produce their own. A very wide range of radioactive materials is available commercially, in the form of elemental radionuclides or as chemical compounds with a radioisotope at a specific point in the molecular structure. One of the world's largest commercial suppliers of radionuclides is the Radiochemical Centre, Amersham, UK, which lists over a hundred radionuclides in routine production with many more available on special order.

The radionuclides in greatest demand by research laboratories are those used to label organic molecules, particularly with the only readily available isotopes of carbon and hydrogen, ^{14}C and ^{3}H respectively. Several hundred organic compounds labelled in specific positions with ^{14}C or ^{3}H are available from the Radiochemical Centre, and preparations to order can be arranged to meet special requirements. Most labelled organic compounds are supplied in quantities which range from a few kBq to many MBq (or μCi to mCi in the older units) and most compounds are supplied at very high specific activities.

Several companies list a range of radionuclides and labelled compounds among their products, and the names of a number of these suppliers are given in the appendix. Note that governments in most countries now require persons keeping or using radioactive materials to be registered with the appropriate authority and to observe certain legal, health and safety regulations.

Bibliography

Friedander, G., Kennedy, J. W. and Miller, J., *Nuclear and Radiochemistry*, J., Wiley & Sons, New York (1964)

McKay, H. A. C., *Principles of Radiochemistry*, Butterworths, London (1971)

Haissinsky, M., *Nuclear Chemistry and its Applications*, Addison-Wesley, Reading MA (1964)

Harvey, B. G., *Introduction to Nuclear Physics and Chemistry*, Prentice-Hall, Englewood Cliffs (1962)

Manual of Radioisotope Production, Technical Reports Series No. 63, International Atomic Energy Agency, Vienna (1966)

3 Detection and Measurement of Radioactivity

3.1 Interaction of radiation with matter

Radioactivity is the spontaneous decay of unstable nuclei and as we have seen this is usually accompanied by the emission of charged particles or photons. The more important forms of radioactive decay are collected in *Table 3.1*, where the emissions associated with the different decay processes are indicated.

The measurement of radioactivity is based largely on detecting the emissions from decaying nuclei, and it is therefore important to understand the interactions of these emissions with other materials. Radioactive decays are generally rather energetic events and the emitted particles or photons have energies which are high compared with the bond strengths and ionisation energies of molecules of matter. α particles are emitted with specific kinetic energies in the range 1.8–11.7 MeV; a typical α particle energy distribution for an α decay radionuclide is shown in *Figure 3.1a.* β^+ and β^- decays on the other hand produce e^+ and e^- particles with a distribution of energies ranging from zero to a maximum value, as shown in *Figure 3.1b*. The maximum β particle energy varies with the radionuclide and ranges from about 18 keV (for β^- from ^3H) to 16.4 MeV (for β^+ from ^{12}N). γ photons emitted from excited state nuclides have discrete energies

Table 3.1 Radioactive decay modes and associated emissions

Mode of decay	*Symbol*	*Emission*
α decay	α	α particle
β^- decay	β^-	e^-
β^+ decay	β^+	e^+
Electron capture	e.c.	X ray
γ emission	γ	γ photon
Isomeric transition	I.T.	γ photon
Spontaneous fission	s.f.	Fission products and neutrons

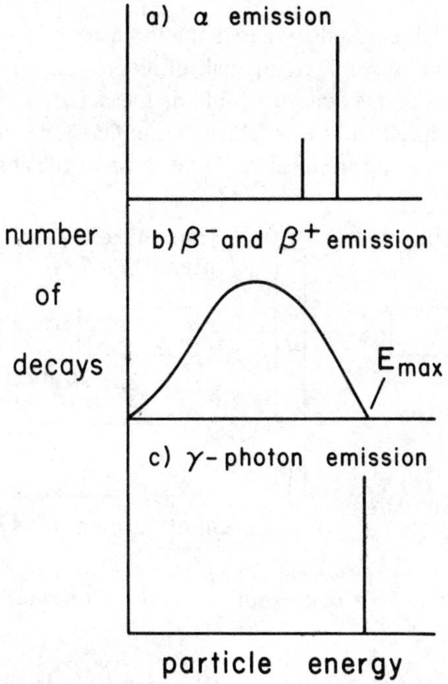

Figure 3.1 Schematic representation of energy distribution of emitted particles following: a, α decay; b, β^- and β^+ decays; c, γ-photon emission.

in the range 0–8.9 MeV, a range which covers the characteristic X rays (0–0.15 MeV) emitted following electron capture decay. A typical γ photon energy distribution spectrum is shown in *Figure 3.1c*.

The energetic particles and photons emitted during radioactive decay interact with matter primarily by causing ionisation, although the way in which this effect can be utilised for the detection of the radiation depends on the state of the matter involved (solid, liquid or gas) and on the efficiency of the interaction. α particles, with their double charge and relatively large mass, lose energy very rapidly as they pass through matter, leaving short straight trails of densely ionised material in their wake. β particles, with a single unit of charge and a low mass, are more readily deflected by collision with molecules and are less efficient than α particles at causing ionisation. Consequently as β particles pass through matter erratic trails of ionised molecules are formed. γ photons, being uncharged and of zero rest mass, interact with molecules rather infrequently, leaving trails through matter which are long compared with those left by the charged particle emissions, and which consist of widely separated ionisation events.

3.2 Mean range

These considerations can be discussed in a more quantitative manner by defining the mean range of a radiation. The mean range for a monoenergetic radiation is the amount of material required to reduce the intensity of the radiation (i.e. particles or photons per second) to one half of its original value. The arrangement in *Figure 3.2* shows how this quantity could be measured, where the thickness of the absorber can be varied and the thickness required to produce $I_x = \frac{1}{2}I_0$ can be determined. The units of mean range could be those of the absorber thickness x,

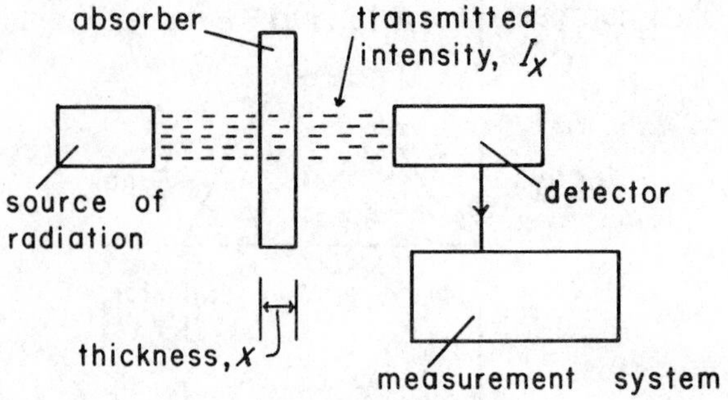

Figure 3.2 Technique for measurement of transmitted intensity of radiation through an absorber of thickness, x. There is no significance in the fact that the radiation depicted is collimated.

e.g. metres, although this would result in the mean range having different numerical values for different absorber materials. If the absorber thickness, x, is multiplied by the absorber density, ρ, we obtain the quantity of matter per unit surface area of the material, e.g. kg m^{-2}. The mean range expressed in these units is, to a first approximation, independent of the nature of the absorbing material. For example, the mean range of the 4.19-MeV α particles emitted during ^{238}U decay is about 0.03 kg m^{-2}, which corresponds to a few millimetres

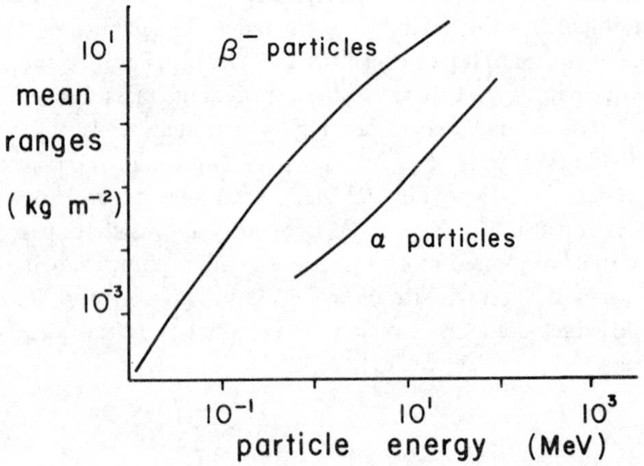

Figure 3.3 Variation of mean ranges for α and β⁻ particles passing through matter with initial particle energy

of air at STP or a single sheet of tissue paper. In practice the mean range of charged particles does vary somewhat with the atomic number of the absorbing material (details are given in the standard texts on nuclear and radiochemistry).

The mean ranges of the charged particle radiations vary dramatically with the initial energy of the radiation. These variations are known as the range–energy relationships and are shown on a log–log scale in *Figure 3.3*. Clearly the range of β particles is more than an order of magnitude greater than that of α particles with the same initial energy. Thus the mean range of 4-MeV electrons is $\sim$0.2 kg m^{-2}, which corresponds to almost a millimetre thickness of aluminium foil.

3.3 γ-photon interactions with matter

The mean ranges of γ photons and characteristic X rays are more complex functions of energy. This complexity arises because these electromagnetic radiations lose energy by two different processes, (and in fact a third which we consider later). The first process is the photoelectric effect, in which the energy of a γ photon is entirely converted into the kinetic energy of an electron ejected from a molecule. The kinetic energy of the resulting photoelectron becomes equal to the γ photon energy less the binding energy with which the electron was held within the electronic structure of the molecule. Photoelectron production can

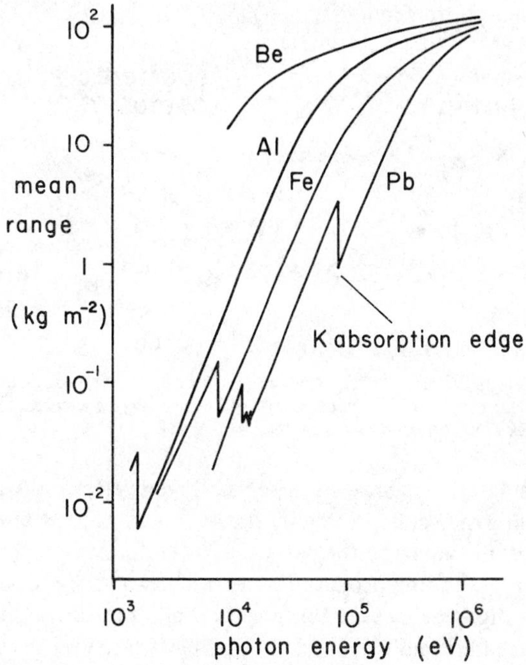

Figure 3.4 Variation of mean ranges for γ photons passing through different materials with initial photon energy

only occur when the energy of the incident γ photon is greater than the binding energy of the electron, and, of course, molecules possess many electrons with different binding energies. For this reason the probability of photoelectron production as a function of γ-photon energy shows sharp discontinuities at photon energies equal to the electron energy levels within the molecule, and these discontinuities occur at different energies for different absorbing materials. This is reflected in the range-energy relationships for γ photons shown in *Figure 3.4* for several absorbing materials. The sharp discontinuities are called absorption edges and are apparent over the energy region occupied by atomic electrons (< 100 keV). The edges are identified by the symbol of the electron shell associated with the discontinuity, such as a K absorption edge.

The second mechanism by which γ photons lose energy is Compton scattering, in which only a part of the photon's energy is converted into kinetic energy of an electron—the rest remaining in the form of a γ photon of lower energy than the incident photon. Compton scattering is shown in *Figure 3.5*. The scattered photon may undergo a second interaction with a different molecule, producing either a photoelectron or a Compton electron and another scattered photon. The probability of Compton scattering and the probability of photoelectron production both decrease as γ-photon energy increases although, as the photoelectric effect probability declines more rapidly with increasing energy than the Compton scattering probability, this latter process becomes a more important mechanism of energy loss at higher photon energies.

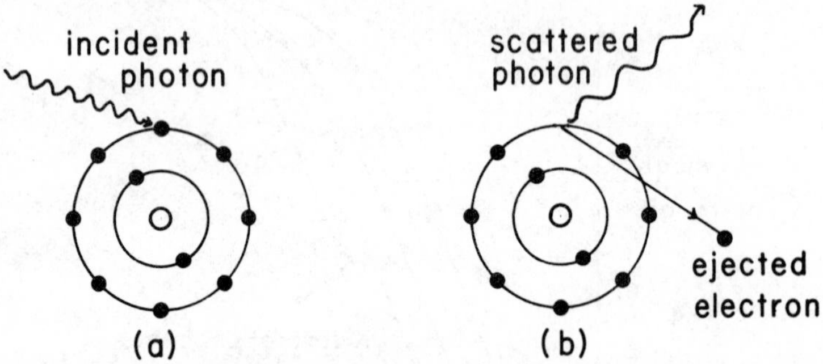

Figure 3.5 The effects of Compton scattering. a, A γ photon interacting with an extranuclear electron of the target atom. b, The departure of the lower energy scattered photon and the ejection of the struck electron after the interaction

Both processes described above give rise to energetic electrons which can interact with surrounding molecules in exactly the same way as β^- particles. In fact most of the ionisation caused by the passage of γ-photons through matter results from the secondary ionisation produced by photoelectrons and Compton electrons. The overall range-energy relationships for γ photons shown in *Figure 3.4* indicate that the mean ranges of γ radiations in matter are very much greater than those of α or β particles of the same energy. Thus a 4 MeV γ photon has a range of ~ 150 kg m^{-2} in lead, corresponding to a thickness of ~ 1.5 cm.

3.4 Gas ionisation detectors

While gas ionisation detectors come in many shapes and sizes, the essential elements of the majority can be understood from a consideration of the device shown in *Figure 3.6*. The detector consists of a metal tube with a thin gas-tight 'window' of mica or mylar at one end and a gas-tight insulated support for the thin central wire at the other end. The tube is filled with a gas mixture, such as

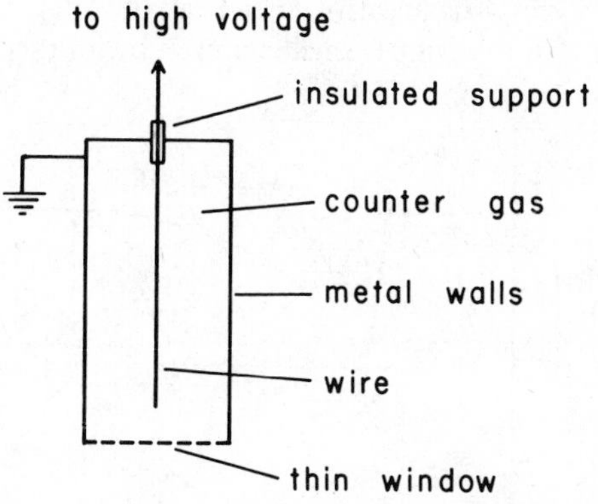

Figure 3.6 Schematic arrangement of a typical gas ionisation detector. In this case the radiation may enter the counter gas through the thin window at the end of the detector

90% argon and 10% methane, which becomes partially ionised when radiation enters the detector. In practice, the ionising events result in the formation of positive ions and free electrons, each pair being called an ion-pair. For many gases the energy required to produce an ion pair averages to about 35 eV, so that, for example, if a 1 MeV β^- particle enters the detector about $10^6/35 = 2.86 \times 10^4$ ion pairs are formed. When the central wire is maintained at a high positive potential with respect to the wall the electron of each ion pair moves rapidly towards the wire, while the positive ions drift slowly towards the wall. Subsequent events depend on the magnitude of the potential difference between the wire and the wall.

For relatively low operating voltages (of the order of 2–400 V for a 2 cm diameter tube) the electrons are simply collected on the wire. As the device has capacity (in the electronic sense) between the wire and the wall, the arrival of negative charge at the wire causes a drop in the wire potential of

$$\Delta V = \frac{n.e.}{C} \text{ V}$$

where n is the number of electrons collected, e is the electronic charge in Coulombs, and C is the detector capacity in Farads.

The drop in wire potential will cause a current to flow from the power supply until the wire voltage has been restored to the supply level, a process which takes a time of the order of

$$t = RC \text{ s}$$

where C is again the capacity of the detector, and R is the resistance in ohms of the circuit connecting the power supply to the wire.

Figure 3.7 shows how the wire voltage varies as a function of time after a β^- particle enters the detector. The negative going peak of voltage is described as a pulse, and, if the width of the pulse is small compared with the time between

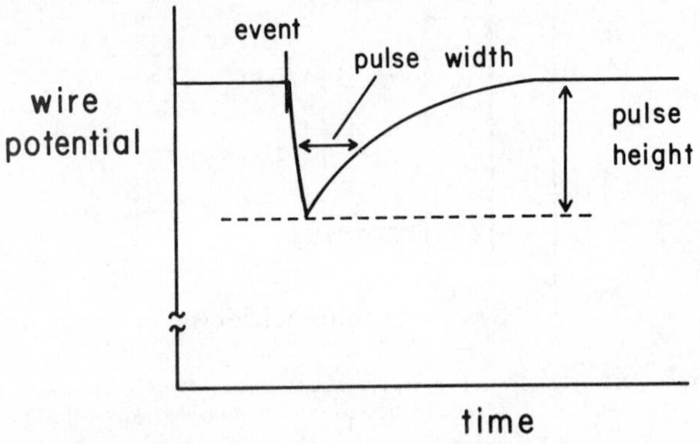

Figure 3.7 Typical variation of wire potential of a gas ionisation detector with time following the interaction of a particle of radiation with the counter gas

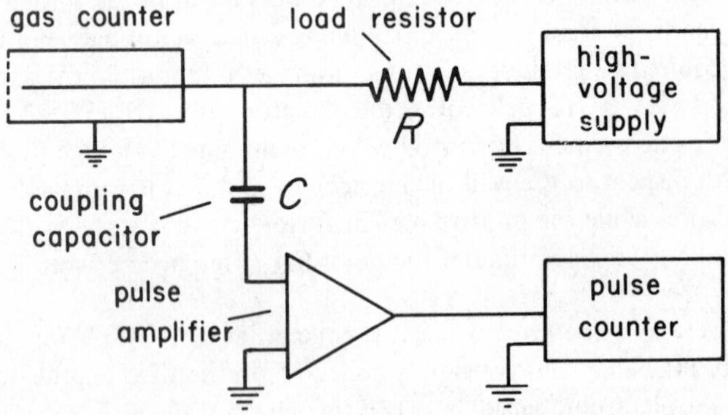

Figure 3.8 Electronic circuit elements associated with a gas ionisation detector for the counting of radiation interaction events within the detector

successive β^- particles entering the detector, counting these pulses provides a simple method of measuring the number of β^- particles being detected. The circuit elements required to operate the detector in this manner are shown in *Figure 3.8*. The resistance, R, is called the load resistor and is usually chosen to be ~ 100 kΩ. The capacity of a typical detector is of the order of 100 pF, so that the width of pulses produced on the wire is ~10 μs. This arrangement could, in principle, be used to count ionising events at a rate up to 20000 s^{-1} before the overlapping of successive pulses becomes a problem.

When operated in the manner outlined above the detector is called an ion chamber. It suffers from one serious disadvantage in that the magnitude of the voltage pulses produced is very small (of the order of microvolts), so that the electronics required to amplify the pulse are both complex and expensive. Fortunately there is a relatively simple solution to this problem.

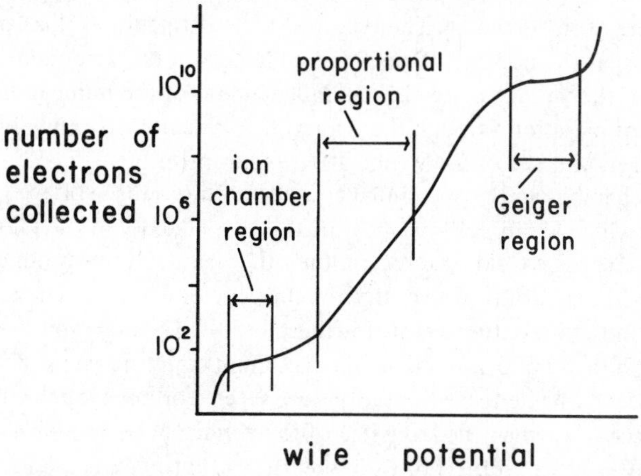

Figure 3.9 Schematic variation of number of electrons collected on the wire of a gas ionisation detector following a radiation interaction event within the detector with the wire potential. Electron numbers shown are typical, although the actual variation would depend on the nature of the counter gas and the nature of the radiation interaction event

If one plots a graph of the number of electrons collected on the wire versus the wire potential, the result is approximately as shown in *Figure 3.9*. As the potential difference between the wire and the wall is increased, so the electrons gain sufficient energy before reaching the wire to cause further ionisation as they collide with gas molecules. Thus the number of electrons collected on the wire becomes greater than (although proportional to) the number created by the passage of radiation through the detector gas. This increase is known as gas amplification. As the wire voltage is increased further the extent of gas amplification increases, and over a range of wire voltages the number of electrons collected varies almost linearly with applied voltage. This range of wire voltage is known as the proportional region, and a detector operating in this region is called a proportional counter. With a suitable choice of counter gas the gas amplification within the proportional region may be as high as 10^7, so that the magnitude of voltage pulses produced on the wire can be much greater when the detector is used as a proportional counter than when operated as an ion-chamber. Therefore much simpler electronic circuitry is required for amplifying pulses from proportional counters and the whole system becomes fairly inexpensive.

At even higher wire potentials the gas amplification increases rapidly (*see Figure 3.9*), reaching a finite limit (the Geiger region) only when the positive ions formed along with the electrons become so numerous that the strength of the electric field between the wire and the wall falls and electrons cease to be attracted to the wire. In this state the detector is said to be saturated; a maximum number of electrons have been collected on the wire and the detector has become insensitive to the action of further radiation. The detector remains in this saturated state until the positive ions have had time to drift to the wall and become neutralised. Since the detector is insensitive during this time, the period is called the dead-time of the detector. The feature which distinguishes this operating condition is that the number of electrons collected on the wire—and so the magnitude of the voltage pulse—is now independent of the number created by the passage of radiation through the detector. A detector operating in this way is called a Geiger-Muller (or G–M) tube after its inventors.

The ion-chamber and proportional counter both produce voltage pulses of magnitudes which are directly proportional to the energies of the radiation entering the detector. The G–M counter, on the other hand, always produces voltage pulses of fixed magnitude, irrespective of the type or energy of the radiation causing the initial ionisation event. Furthermore the G–M counter has a dead time of between 200 and 500 μs, so that the maximum counting rate is limited to less than $\sim 1000\ s^{-1}$. While these limitations are severe for many applications there are other occasions when the large size of the voltage pulse produced by a G–M tube (1–10 V) can be useful. For example, the output pulse is usually large enough to obviate the need for any pulse amplification—a simple counter being the only item of electronics required.

Gas ionisation detectors provide a relatively inexpensive means of detecting radioactivity. Sealed tubes are available from many major component manufacturers, such as Mullard Ltd in the UK, and can often be supplied with a choice of window thickness or position. Most have a suggested operating voltage and G–M tubes usually have their dead times quoted in a specification sheet. Electronic amplifiers, digital counters, count rate meters and high-voltage supplies are available as modular units or integrated into a single unit from nuclear equipment manufacturers such as Nuclear Enterprises Ltd and Panax Equipment Ltd.

Gas ionisation detectors have the advantage of being simple and robust items of equipment. Unfortunately they have the disadvantage of having a window which, however thin, prevents low energy α and β radiation from entering the detector with high efficiency. In fact ^3H, with its maximum β^- energy of 18 keV, cannot be efficiently detected by a counter fitted with the thinnest windows available, and other low energy β^- emitters such as ^{14}C and ^{35}S are counted with efficiencies of only a few per cent by window counters. This problem can be overcome to some extent by using open-ended or windowless counters, through which counter gas is passed from a cylinder. Windowless counters allow solid samples to be placed virtually inside the counter gas, and this certainly results in increased counting efficiency for weak α and β emitters—or at least for those radiations which escape from the surface of the sample. However, the greater complexity of the system, involving control of gas flow, and the dependence of the counting efficiency on the purity of the counter gas has resulted in the increased use of alternative techniques for the detection of these weak radiations.

A second disadvantage of any gas-filled counter stems from the fact that the gas is the 'target' for the radiation. The large range associated with energetic γ photons results in very poor detection efficiencies for γ radiation above $\sim$100 keV. For example, a typical G–M counter may detect the 662 keV γ photons from ^{137m}Ba with an efficiency of $<0.1\%$; the 99.9% of undetected γ photons passing straight through the counter without causing ionisation.

For these reasons gas ionisation detectors are used mainly for the detection of radioactivity from energetic α or β emitting isotopes such as ^{32}P (maximum β^- energy $\sim$1.3 MeV), the low energy ($<$25 keV) γ emitters or X ray emissions from some isotopes which decay by electron capture. Other emissions are more efficiently detected using one of the techniques described below.

3.5 γ-scintillation detectors

Probably the most widely used technique for detecting γ radiation is based on the fact that when energetic photons enter certain crystals, short duration flashes of visible light, known as scintillations, are observed. Crystals of several organic and inorganic compounds exhibit this behaviour, although by far the most popular crystal for γ-photon detection is sodium iodide containing small quantities of thallous iodide, written NaI(Tl). Sodium iodide crystals of this kind are available commercially, usually grown in cylindrical form and in a variety of sizes up to more than 15 cm in diameter. The crystals normally found in the radiochemistry laboratory are likely to be 5–10 cm in diameter, and even these may cost several hundred pounds. With the aid of a scintillator crystal the problem of detecting γ photons reduces to the problem of detecting visible light scintillations.

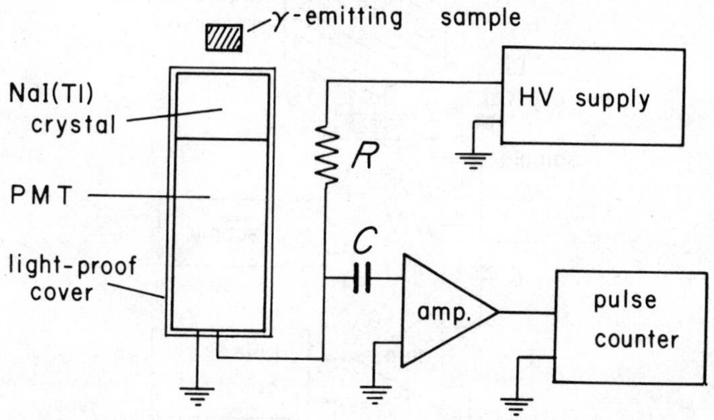

Figure 3.10 Schematic arrangement of a crystal scintillation detector for detecting γ photons. Associated electronic circuit elements shown are required for counting the number of γ photons detected

The standard device for detecting small numbers of visible light photons is called a photomultiplier tube (PMT). Essentially it converts light flashes into electrical pulses, and so produces an output which is similar to that of the gas proportional counter. The conventional arrangement for detecting γ photons using an NaI(T1) crystal scintillator is shown in *Figure 3.10*. Because the photomultiplier tube is sensitive to light, the tube and the scintillator crystal are enclosed in a light-proof cylinder, the end cap over the crystal being sufficiently thin to allow γ radiation to enter the crystal without significant attenuation. When a γ photon interacts with the crystal a scintillation, consisting of several visible light photons, is detected by the PMT and converted into a shower of electrons which are collected on the positive electrode (the anode) of the tube. As with the gas ionisation detectors this produces a negative going pulse of short duration ($< 1\ \mu s$ is typical) on the anode, and these pulses are amplified and passed to an electronic pulse counter. As with the proportional counter it is observed that the magnitude of voltage pulses from the scintillation detector can be proportional to the energy of the γ photons giving rise to the scintillations. The applications of this proportionality will be considered in a later chapter.

3.6 Semiconductor detectors

Since the 1960s a new type of device has become increasingly popular for detecting γ radiation. This is the semiconductor detector illustrated in *Figure 3.11* (full details of the operating principles are given in more advanced texts). A simple idea of the operation of a semiconductor detector may be obtained by recalling that in appropriate bias conditions a semiconductor junction may act as an insulator. When the passage of radiation gives rise to ionisation within the junction

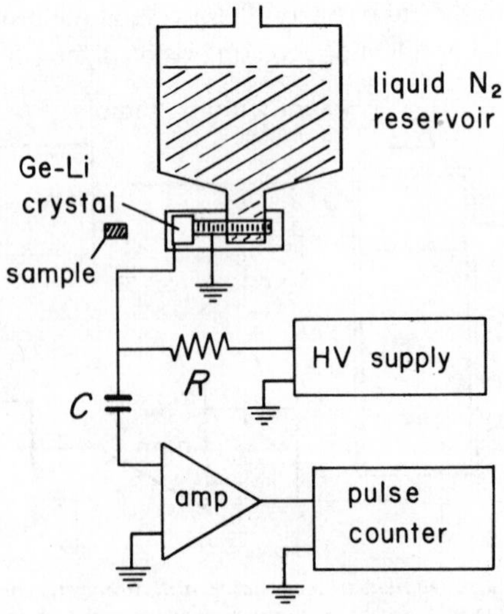

Figure 3.11 Schematic arrangement of a typical semiconductor detector for γ photons

the electrons released will travel to the positively biased side of the junction, constituting a momentary flow of charge, i.e. a current. It is the charge released in this way that is collected and amplified into a current pulse for counting.

Semiconductor detectors fall roughly into two types, depending on the nature of the semiconductor material. One type consists of very pure semiconductor material, such as germanium, in which case the device is called an intrinsic germanium detector. The second type consists of less highly purified semiconductor material into which a second element has been diffused or 'drifted'. The most widely used example of this class of detector is the lithium drifted germanium detector, otherwise known as the Ge–Li (pronounced jelly) detector. The lithium atoms react with impurities in the germanium and carry the impurities through the crystal as the drifting occurs, leaving behind a region of very pure germanium of suitable conductivity for γ photon detection. Owing to the mobility of the small lithium atoms within the germanium lattice Ge–Li detectors must be maintained at liquid nitrogen temperatures (77 K) at all times—whether in use or not—otherwise the lithium so expensively drifted into the semiconductor crystal drifts out again. Both types of germanium semiconductor detector are operated at liquid nitrogen temperatures, although some newer semiconductor detectors based on silicon have recently become available for use at higher temperatures.

Although semiconductor detectors have considerable advantages in γ spectrometry (*see* chapter 6), for many radiochemical applications their disadvantages become rather serious. The principal disadvantage lies in the fact that the semiconductor crystals are small, with lattices of relatively low mass atoms. This results in relatively low detection efficiences for γ photons, particularly for photons with energies greater than a few hundred keV. This in turn can result in rather poor sensitivity for the detection of γ-emitting isotopes, requiring longer counting times than would be necessary with a crystal scintillation detector. A second disadvantage is that the width of the electrical pulses produced by semiconductor detectors is usually much greater than for the pulses produced by scintillation detectors, pulse widths of $\sim 100\,\mu s$ being common for germanium based detectors, and this in turn limits the counting rate. This problem can be overcome electronically to some extent, although the additional electronic equipment required adds to the cost of an already expensive detection system.

3.7 Detection efficiency

The three main types of detector outlined above, gas ionisation, scintillation and semiconductor, can in principle be used for detecting α, β or γ radiation. However, the wide variations in detection efficiency with both the nature and energy of the radiation limit the choice of detection system for particular applications. For α particles a window-type gas ionisation counter is of little value because of the absorption of α particles in the window material. Scintillation counters can be used for α particles, provided the light-proof cover around the scintillator is very thin. In practice zinc sulphide containing small amounts of silver, ZnS(Ag), is used as the scintillator material for α particle detection, usually in the form of a thin layer coated onto a plastic disc which itself is in contact with the PMT. For samples which can be brought very close to the scintillator fairly respectable counting efficiencies can be obtained, say 10%, although absorption by the sample material or by air makes it difficult to give reliable numerical values. Semiconductor detectors are good for α particles although as before only a very thin detection area is required, the variety normally used being called a surface barrier detector. Again the very small mean range characteristic of even energetic α particles makes the actual detection efficiency highly dependent on the sample thickness and distance in air between the sample and detector.

For β^- emitting nuclides the gas ionisation detectors can provide quite acceptable counting efficiencies for all but the weakest emitters (e.g. ^{3}H, ^{14}C, ^{35}S, ^{63}Ni), and even these can be counted using windowless detectors. For example, a typical laboratory G–M counter will count ^{32}P (β^- energy max = 1.7 MeV) with an efficiency of ~30%. Scintillation counters may also be used for energetic β emitters, although the scintillator material is normally a few millimetres of a plastic scintillator with a light proof coating, rather than NaI(T1) which requires a thicker covering to keep out water vapour as well as light. The overall counting efficiency for β^- emitters of a plastic scintillation counter may be slightly higher than that of a gas ionisation detector, as the light tight coating on the scintillator may be thinner than the window of a gas counter, but for energetic β^- emitters the difference between the two is negligible. Semiconductor detectors are rarely used for counting β^- emitters simply because the alternatives are much cheaper and just as efficient.

Low energy γ emitting isotopes can be detected using gas ionisation detectors, particularly detectors filled with high atomic number counting gases such as Kr and Xe. Indeed for X ray emitters (e.g. the electron capture decay nuclides) a gas

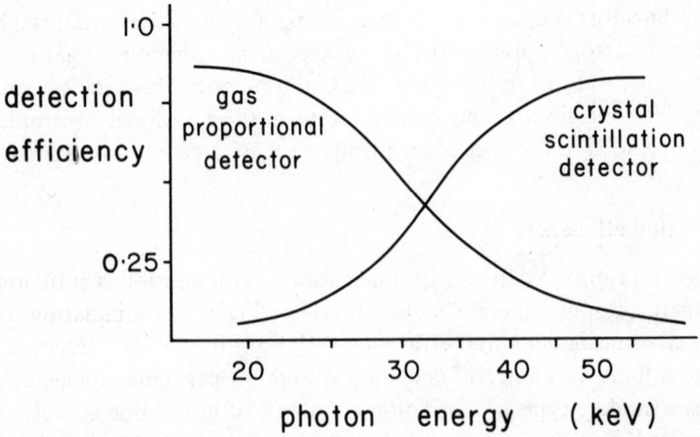

Figure 3.12 Variation in detection efficiency with photon energy for typical gas proportional detector and crystal scintillation detector when used for detecting X ray or low-energy γ photons. Actual variation would depend on detector size and composition

ionisation detector may be more efficient than a crystal scintillation detector. *Figure 3.12* shows the variation in detection efficiency for low energy γ (or X) radiation of a typical proportional counter and a typical NaI(T1) crystal scintillation counter. At energies below 30 keV the proportional counter would clearly be the preferred detector, while at higher energies, say about 50 keV, the NaI(T1) detector is much better. For high-energy γ photons (above, say 500 keV) gas filled detectors are virtually useless having detection efficiencies of $< 0.1\%$. Crystal scintillation counters are excellent in the energy range between 50 keV and 2 MeV but at higher energies these also have lowered detection efficiencies. To some extent the decline of counting efficiency with increasing photon energy can be offset by increasing the size of the detecting crystal, although this rapidly becomes a very expensive way of achieving a rather small increase in efficiency. As already mentioned, most semiconductor detectors have rather low counting efficiencies at best, and above the 100 keV region cannot be regarded as serious competitors for the crystal scintillation detectors.

3.8 Position annihilation

Energetic positrons interact with matter in much the same way as energetic electrons, and we have therefore considered the detection of β^+ and β^- decay nuclides as a single problem. However, once a positron has lost its kinetic energy it succumbs to the attraction of an electron and collides with it. This meeting of oppositely charged electrons results in the annihilation of both, the vanished mass appearing in another form of energy as two 0.511 MeV photons travelling in directly opposing directions. These two photons are called annihilation radiation and may be detected in just the same way as γ photons of similar energy.

Thus a β^+ decay isotope may be detected either by its charged particle emission or by the annihilation radition which ultimately leaves the sample or the surround-area. In practice, for solid or liquid samples the detection of the annihilation radiation is most convenient because this presents no self-absorption problems, and because the photon energy of 0.511 MeV may be detected with high efficiency ($> 60\%$) using a simple NaI(T1) crystal scintillation detector.

3.9 Detector shielding

One of the problems associated with the detection of radioactivity is that the surface of the Earth is bathed with radiation from sources which we do not normally wish to detect. For example, cosmic radiation and radioisotopes in our environment give rise to a flux of electromagnetic radiation which is present in every laboratory in the world. Most of this radiation is in the energy range below 0.5 MeV, so that most detectors can detect it. Semiconductor detectors have rather small detection volumes so that this background radiation can usually be ignored. However, G-M tubes have a volume of, typically, 20 cm^3 and in an open laboratory these (and the other gas ionisation detectors) will record of the order of 50 counts per minute from background radiation. Crystal scintillation counters vary considerably in

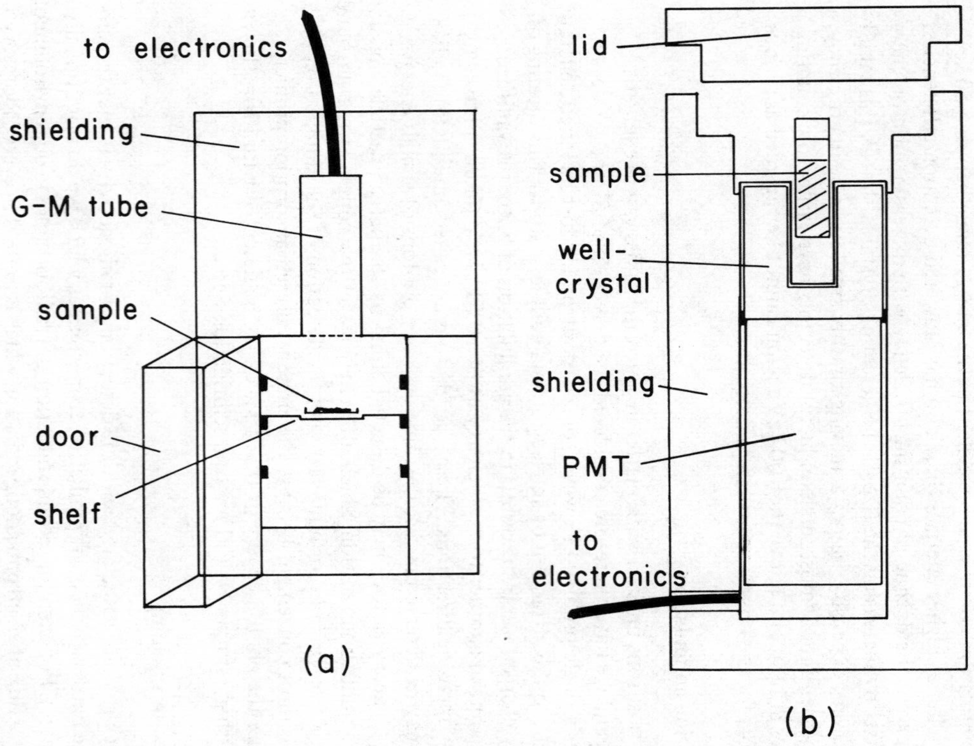

Figure 3.13 Schematic representation of typical lead shielding 'castles' used to house: a, a G–M end-window detector; and b, a well crystal γ-scintillation detector

volume but a 7.5 cm diameter crystal will normally allow several hundred counts per minute to be recorded from background radiation and noise from the photomultiplier tube and amplifier electronics.

To minimise the effects of background radiation these types of counters are almost invariably shielded by several centimetres of lead. This lead shielding absorbs a significant fraction of the background radiation and so prevents it reaching the detector. While shielding may be constructed from lead bricks, most radiochemistry laboratories are equipped with purpose built lead castles designed to shield particular types of detector. Thus a G–M counter may be housed in a castle such as that shown in *Figure 3.13a*, in which the tube faces down so that solid samples may be placed inside the castle and directly under the counter window. On the other hand, a γ-scintillation counter may be held in a castle so that the detector faces upwards, as shown for a typical well-crystal scintillation counter in *Figure 3.13b*. In this case the sample is contained in a small vial (usually $\sim$5 ml volume) which is fitted into the well of the detector, and a lead lid positioned above the crystal.

While background radiation cannot be completely eliminated by shielding it should be possible to operate gas counters with background count rates of < 10 min^{-1}, and 5.0 cm crystal scintillation counters with background count rates of < 30 min^{-1}.

Bibliography

Friedlander, G., Kennedy, J. W. and Miller, J., *Nuclear and Radiochemistry*, J. Wiley & Sons, New York (1964)

McKay, H. A. C., *Principles of Radiochemistry*, Butterworths, London (1971)

Haissinsky, M., *Nuclear Chemistry and its Applications*, Addison-Wesley, Reading MA (1964)

Overman, R.T. and Clark, H. M., *Radioisotope Techniques*, McGraw-Hill, New York (1960)

Nicholson, P. W., *Nuclear Electronics*, J. Wiley & Sons, New York (1974)

4 Statistics of Counting

4.1 The arithmetic mean

Most experiments which involve radioisotopes require the quantitative estimation of radioactivity based on the number of counts recorded by some detecting apparatus during a given time interval. If one takes a sample of radioactive material (with a half-life which is long compared with the time scale of this experiment) and counts the number of decays detected in a series of equal time intervals, statistical fluctuations in the decay rate will produce different numerical values for most of the recorded results. For example, *Table 4.1* shows the results of one such experiment where a radioactive sample was counted for 10 one-minute periods and the following results were obtained:

Table 4.1 Counts per minute obtained from a radioactive sample

Minute	*Count*
1	89
2	120
3	94
4	110
5	105
6	108
7	85
8	83
9	101
10	95

As all the values are different one must consider which value, or what other number, most accurately reflects the decay rate of the sample.

For N separate counting periods in which counts of $x_1, x_2, x_3 \ldots x_N$ are recorded, we may define the arithmetic mean of the result, $\bar{x}$, by

$$\bar{x} = \frac{1}{N} \sum_{i=1}^{N} x_i \tag{4.1}$$

For the sequence given in *Table 4.1* the arithmetic mean is 99. However, if we repeat the experiment and obtain another 10 values for the count, the chances are that we would obtain a different arithmetic mean. In fact a 'true' arithmetic mean would only be obtained by taking an infinite number of counts of the sample. As time does not usually allow this, we must make an estimate of true arithmetic mean. For the data given above our best estimate of the true arithmetic mean must be 99. But how close is this likely to be to the true arithmetic mean? We will attempt to answer this question quantitatively below.

4.2 Distribution functions

If a very large number of counting experiments are performed one could plot a distribution of the number of times, n_i, that a count, x_i, is recorded. This distribution would have the general form shown in *Figure 4.1*. The curve is a graphical representation of the distribution function of the data, and for radioactive decay the function is called a Poisson Distribution.

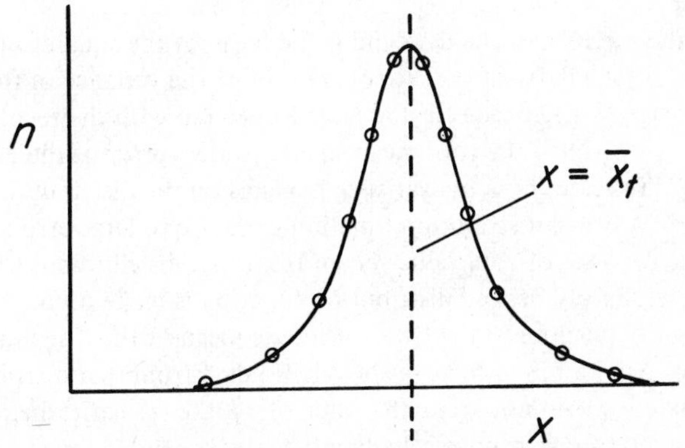

Figure 4.1 Distribution function showing the number of times, n, that the result of counting a given sample for a fixed time is x

If there are a sufficiently large number of experiments then the mean of all the results will be $\bar{x}_t$, the true arithmetic mean. For the result x_i we may define an error

$$\Delta_i = x_i - \bar{x}_t$$

and clearly Δ_i represents the error in the result x_i, in the sense that x_i differs from $\bar{x}_t$ by Δ_i. A measure of the precision of N data values scattered about the true mean can be obtained from

$$\sigma_t^2 = \frac{1}{N} \sum_{i=1}^{N} \Delta_i^2 = \frac{1}{N} \sum_{i=1}^{N} (x_i - \bar{x}_t)^2 \qquad (4.2)$$

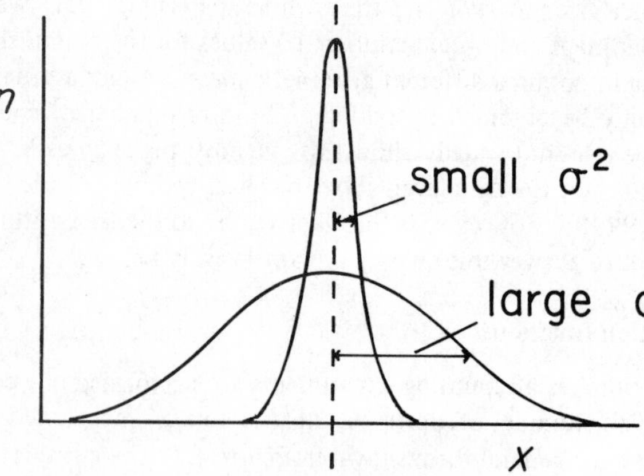

Figure 4.2 Effect of the magnitude of the variance, σ^2, on the shape of the distribution function

σ_t^2 is called the variance of the data and is the mean of the squares of the error associated with each data value. The square root of the variance of the data, σ_t, provides an average magnitude for the error associated with the recorded x_i values; σ_t is often called the root mean square (r.m.s.) error of the data values.

Figure 4.2 illustrates how the variance depends on the distribution function of data values. A broad distribution function gives rise to large errors and so to a large variance for a set of data values taken from that distribution. On the other hand a narrow, sharply peaked distribution function is made up of data values which frequently lie close to the true arithmetic mean of the distribution, so a small variance will be found for a set of values taken from that distribution.

Unfortunately we do not know the value of $\bar{x}_t$, the true arithmetic mean of a distribution, and so we cannot calculate the true variance of a set of data values from equation 4.2. We do have an estimate of the arithmetic mean, $\bar{x}$, for the data, and can obtain an estimated variance of the data, σ^2, using a somewhat modified definition:

$$\sigma^2 = \frac{1}{N-1} \sum_{i=1}^{N} (x_i - \bar{x})^2 \qquad (4.3)$$

For the counts given in section 4.1 it is a simple matter to show that $\sigma^2 = 142$; and so, $\sigma \triangleq \pm 12$. Hence each of the recorded counts has an average error or standard deviation of ± 12.

4.3 Chauvenet's criterion

In a series of counts one or two values may differ from the mean by so much that doubt is cast on the validity of these particular values. In counting radioisotopes this situation occasionally arises when electrical interference is picked up by the

Table 4.2 d_N values for rejection of suspect data point using Chavenet's criterion

Number of observations (N)	d_N
4	1.53
5	1.64
6	1.73
7	1.80
8	1.87
9	1.91
10	1.97
20	2.24
30	2.39
40	2.50
50	2.57
100	2.80

amplifiers used in electronic counting apparatus, particularly in laboratories where radio frequency or microwave power is in use. Various criteria have been suggested for deciding whether a suspect data point should be discarded. One of these is Chauvenet's criterion, which calls for the rejection of a data value, x_i, if

$$\frac{|x_i - \bar{x}|}{\sigma} > d_N$$

where d_N has a value which depends on the total number of observations, N. Some of the values for d_N are given in *Table 4.2*. Once a suspect data value has been rejected the mean and variance of the remaining data may be recalculated.

4.4 The variance of the mean

The procedures we have adopted so far have allowed us to estimate the mean of a series of data values, and the r.m.s. error associated with the individual values. While we may take the estimated mean, $\bar{x}$, to be our best estimate of the true mean of the distribution function from which a limited number of data values have been taken, we have not yet considered how close $\bar{x}$ is likely to be to the true mean $\bar{x}_t$.

If we repeated the series of 10 counting experiments discussed in section 4.1 we would be able to derive another estimated mean, which would probably be different from our first estimate of $\bar{x} = 99$. By performing many series of 10 counts a distribution of estimated means could be obtained, which would be characterised by its own variance, the variance of the estimated mean.

For a total number of data values, N, an estimate of the variance of the estimated mean, $\sigma_{\bar{x}}^2$, is given by

$$\sigma_{\bar{x}}^2 = \frac{\sigma^2}{N} = \frac{1}{N(N-1)} \sum_{i=1}^{N} (x_i - \bar{x})^2 \tag{4.4}$$

As before, the square root of this quantity, $\sigma_{\bar{x}}$, represents an average error, associated this time with the estimated mean determined from the N data values.

For our original set of 10 counts

$$\sigma_{\bar{x}}^2 = 14.2 \text{ and } \sigma_{\bar{x}} \simeq 4$$

So that we can write the estimated mean of the counts recorded in one minute intervals as $\bar{x} = 99 \pm 4$.

4.5 Reliability of the estimated mean

The results of counting experiments are usually recorded in the form

$$\bar{x} \pm k\sigma_{\bar{x}}$$

where k is a constant which will determine the magnitude of the error we associate with the estimated mean. The probability, P, that the true mean of the distribution from which the observed counts have been obtained lies between $\bar{x} - k\sigma_{\bar{x}}$ and $\bar{x} + k\sigma_{\bar{x}}$ is shown in *Table 4.3*.

Table 4.3 Probability of the true mean of the distribution lying between $\bar{x} - k\sigma_{\bar{x}}$ and $x + k\sigma_{\bar{x}}$

k	P	*Error range called*
0.6745	0.5	probable error
1.0	0.6827	standard error
1.6449	0.9	reliable error
1.96	0.95	95% limits
2.0	0.9545	2σ limits
2.5758	0.99	99% limits

Thus if a series of counts gives an estimated mean of 200 min^{-1} and $\sigma_{\bar{x}}$ = 10 min^{-1} we could say: (i) the probable error is ± 6.7 min^{-1}, and there is a 50% probability that the true mean count lies between 193.3 and 206.7 min^{-1}; (ii) the standard error is ± 10 min^{-1}, and there is a 68.3% probability that the true mean count lies between 190 and 210 min^{-1}; (iii) the 2σ limits are ± 20 min^{-1}, and there is a 95.5% probability that the true mean count lies between 180 and 220 min^{-1}.

4.6 Error limits for a single count

Estimating the mean of a distribution function and the variance of the estimated mean from a single count may sound rather incongruous. Nevertheless, the statisticians have developed an approximation that is used widely by radiochemists for this purpose. As one has only obtained a single data value, x, it is not surprising to find that the approximation is

$$\bar{x} \simeq x \text{ and } \sigma_{\bar{x}}^2 \simeq x \text{ so that } \sigma_{\bar{x}} \simeq x^{1/2}$$

Thus the result of a single counting experiment is usually recorded as

$$\bar{x} = x \pm kx^{1/2} \tag{4.5}$$

where k has the same significance as discussed above.

Many of the automatic counting systems available commercially are designed to count each radioactive sample just once, and to print out the number of counts recorded in any desired time interval together with the error limits associated with this result. The error limits used frequently are the 2σ limits, so that the result may be printed as

$$x \pm 2x^{1/2}$$

or the percentage error, E, calculated from

$$E = \frac{2x^{1/2}}{x} \times 100\%$$

4.7 Cumulative errors

Most quantitative results from experiments involving radioisotopes are obtained by combining two or more quantities, each of which has its own associated error limits. Even the simplest kind of activity determination, for example, may require the counting of a radioactive sample followed by the counting of the background radiation to which the detector is sensitive—the background radiation arising from cosmic radiation, the natural radioactivity of the ground and so on. As the number of counts recorded for the sample will include a contribution from the background radiation, we must generally subtract the background count rate from that determined for the sample to arrive at the count rate produced by the radioactivity of the sample alone.

To set error limits on a final result we must be able to combine the error limits of the individual quantities which contribute to the result. The basic rules for combining errors are summarised below. If we take two quantities, A and B, each with its own error limits, say $\pm a$ and $\pm b$ respectively, the error limits for the combined quantities are as follows:

$$(A \pm a) + (B \pm b) = (A + B) \pm (a^2 + b^2)^{1/2} \tag{4.6}$$

$$(A \pm a) - (B \pm b) = (A - B) \pm (a^2 + b^2)^{1/2} \tag{4.7}$$

$$(A \pm a) \times (B \pm b) = (A \times B) \pm (a^2 B^2 + A^2 b^2)^{1/2} \tag{4.8}$$

$$\frac{(A \pm a)}{(B \pm b)} = \frac{A}{B} \pm \left(\frac{a^2 B^2 + A^2 b^2}{B^4}\right)^{1/2} \tag{4.9}$$

Not surprisingly the type of error limits combined using these rules should be the same for both quantities: both should be standard errors or both should be 2σ limits etc. The type of the combined error limits is then the same as the type of its constituent parts.

4.8 Simple applications

(1) Counting time required for standard error limits of $E\%$

When counting samples of low activity it is often desirable to know how much counting time is required to achieve a result which will have a particular error limit. Suppose we wish to count a sample for sufficient time, t, to obtain a result, N, with standard error limits of $E\%$. (We will neglect the background radiation in this example.)

We may first perform a count for, say, one minute to obtain an approximate count rate of R min^{-1}.

Now $N \approx R.t$ and the standard error limit for a count of N will be $\pm N^{1/2}$, or in percentage terms

$$\pm \frac{N^{1/2}}{N} \times 100\%$$

We require

$$E = \frac{100}{N^{1/2}}$$

which is

$$E \approx \frac{10^2}{(R.t)^{1/2}}$$

so that

$$t \approx \frac{10^4}{RE^2} \text{ min}$$

(2) Dividing counting time between sample and background counting to obtain minimum error in the true sample counts

The standard method for obtaining the number of counts produced by a radioactive sample is to perform one count for sample plus background radiation, a second count for background radiation alone and subtract the two count rates to obtain a result for the sample. The error limits on the true sample count is thus a function of the error limits for background and sample plus background, and can be minimised by a proper division of available counting time between the two counting periods.

If the approximate count rate for background and sample plus background are first determined to be R_b and R_{S+b} min^{-1} respectively, and the required counting times are t_b and t_{S+b} min^{-1}, then the variance of the estimated mean for the true sample count will be, from equation 4.7

$$\sigma_{\bar{S}}^2 = \frac{R_{S+b}}{t_{S+b}} + \frac{R_b}{t_b} \tag{4.10}$$

and the standard error limits will be $\pm\,\sigma_{\bar{S}}$. We require that $\sigma_{\bar{S}}$ has a minimum value subject to the condition that

$$t_b + t_{S+b} = t_{\text{total}}$$

where t_{total} is the (fixed) total available counting time.

Thus we require

$$\mathrm{d}\sigma_{\bar{S}} = 0$$

while

$$\mathrm{d}(t_b + t_{S+b}) = 0$$

or

$$\mathrm{d}t_b = -\,\mathrm{d}t_{S+b}$$

Differentiating equation 4.10 we obtain

$$2\sigma_{\bar{S}}.\mathrm{d}\sigma_{\bar{S}} = \frac{-R_{S+b}}{t_{S+b}^2}\mathrm{d}t_{S+b} - \frac{R_b}{t_b^2}\,\mathrm{d}t_b$$

which for $\mathrm{d}\sigma_{\bar{S}} = 0$ gives

$$\frac{R_{S+b}}{t_{S+b}^2}\,\mathrm{d}t_b - \frac{R_b}{t_b^2}\,\mathrm{d}t_b = 0$$

hence

$$\frac{t_b}{t_{S+b}} = \left(\frac{R_b}{R_{S+b}}\right)^{1/2}$$

Thus to achieve the minimum error limits the total time available should be divided between background and sample plus background in the ratio given by the square root of the respective (and approximate) counting rates.

Bibliography

Young, H. D., *Statistical Treatment of Experimental Data*, McGraw-Hill, New York (1962)

Topping, J., *Errors of Observation and their Treatment*, Chapman and Hall, London (1972)

Erricker, B. C., *Advanced General Statistics*, Hodder and Stoughton, London (1971)
Kennedy, J. B. and Neville, A. M., *Basic Statistical Methods for Engineers and Scientists*, Thomas Y. Crowell Company, New York (1976)
Walpole, R. E. and Myers, R. H., *Probability and Statistics for Engineers and Scientists*, Macmillan, London (1978)

5 Liquid Scintillation Counting

5.1 The weak radiation problem

Carbon –14 and tritium are β^- emitting radioisotopes with β^- emissions of very low energy which are extremely difficult to detect with any form of window counter, due to self-absorption of the β^- particles and their absorption within the counter window. The window problem has been overcome to some extent by counting solid samples within a windowless G–M or proportional counter, where efficiencies of 15% for ^{14}C and 2.5% for ^{3}H have been reported. To reduce the self-absorption losses it is desirable to mix the active sample homogeneously with the detecting material. This can be done by counting the sample in the gaseous phase. For example, ^{14}C samples may be converted into $^{14}CO_2$ by combustion or by evolution from a labelled carbonate. The gaseous activity can then be intimately mixed with the filling gas of any type of gas ionisation detector, thus minimising the effect of β^- absorption and resulting in high counting efficiencies. Tritium can also be counted in the gaseous state as hydrogen, water vapour or a hydrocarbon.

Since the 1960s a second method of overcoming the β^- absorption problem has become widely used. In this the radioactive sample and a scintillator material are both dissolved in a suitable solvent, and the resulting scintillations are detected and counted. The method is called liquid scintillation counting.

5.2 Outline of liquid scintillation counting

If a compound containing an α or β emitting isotope is dissolved in a solvent such as toluene, the radioactive emissions result in the formation of electronically excited solvent molecules. If the solution also contains a small amount of a suitable scintillator, the excited solvent molecules rapidly transfer their excitation energy to the scintillator, forming electronically excited scintillator molecules, which then relax by the emission of photons. The processes involved are summarised in *Figure 5.1*.

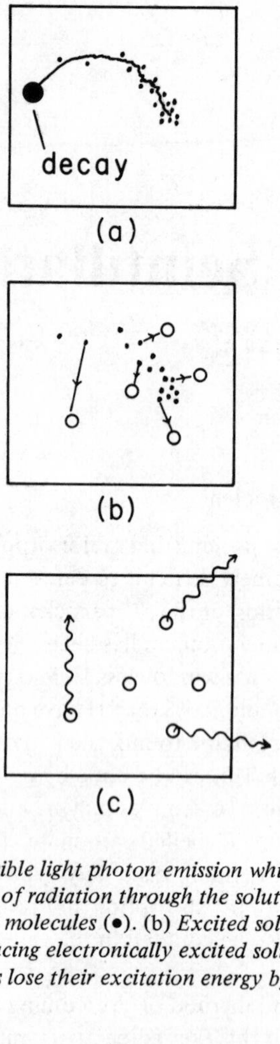

Figure 5.1 The mechanism of visible light photon emission which forms the basis of liquid scintillation counting. (a) *Passage of radiation through the solution produces a trail of ionised and electronically excited solvent molecules* (•). (b) *Excited solvent molecules transfer their energy to solute molecules, producing electronically excited solute molecules* (○). (c) *Electronically excited solute molecules lose their excitation energy by collisional de-excitation or by photon emission*

The scintillator material is usually chosen so that the wavelength of the emitted light is in a region of the spectrum which can be conveniently detected by a photomultiplier tube, e.g. blue light. The result of the combination of radioisotope, solvent and scintillator is that each radioactive decay in the solution gives rise to a flash of visible light, so that electronic counting of these scintillations gives a measure of the activity of the radioactive material. As we shall see later, one of the problems in liquid scintillation counting is that in fact not all the scintillations can be detected, so that the counting efficiency is normally $< 100\%$ and the actual efficiency must be determined before the activity of the radioactive material can be calculated.

While it is possible to arrange for the scintillation to be detected using a single photomultiplier tube, the general noise level of a PMT at room temperature is so high that the separation of the signal pulses (due to scintillations) from the noise pulses (due to thermionic emission from the photocathode of the PMT) becomes difficult. This is especially true for low decay energy isotopes as the β^- emitter ^{3}H, ^{14}C and ^{35}S. Cooling the PMT can help reduce thermionic noise but in fact most commercial liquid scintillation counters use two PMTs, recording a count only when a scintillation is detected by both tubes within a short time period (usually ~ 1 μs). This arrangement is shown in *Figure 5.2*. The technique is called coincidence counting and is highly effective at discriminating against the random noise pulses from the two tubes.

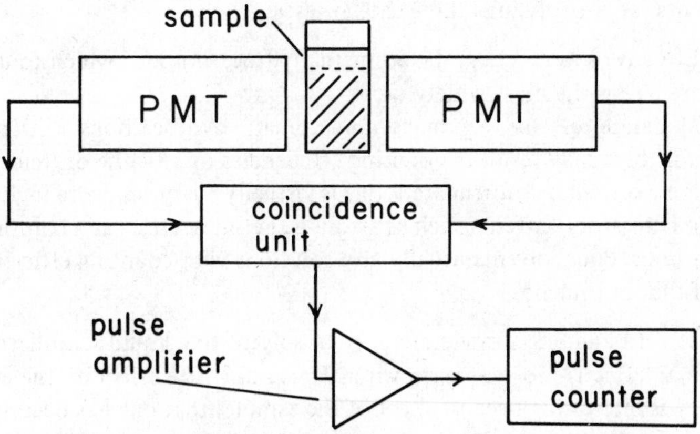

Figure 5.2 Schematic arrangement of photomultiplier tubes (PMTs) and associated electronic units required for coincidence counting in simple liquid scintillation counters. The PMTs and the sample are housed in a light-proof enclosure

The counting efficiency of a liquid scintillation system may be defined as

$$\frac{\text{Number of scintillations per minute detected}}{\text{Number of disintegrations per minute in solution}}$$

Clearly this overall efficiency is made up of two components: one from the scintillator solution itself, and one from the photon detection system and its associated electronics. We will consider these two aspects of the system in turn.

5.3 Liquid scintillator solutions

A wide range of scintillator solutions is available in modern radiochemical laboratories. To make a sensible choice for a particular application, the function of the components of the solution must be understood. All scintillator solutions contain: (1) a solvent; (2) a primary 'solute'–the scintillator material, and may contain:

(3) a secondary solute. The nature and functions of each component are considered below.

(1) The solvent

Despite its name the function of the solvent is not simply to dissolve the radioactive sample. The solvent's functions are to keep the scintillator or solute in solution, and to absorb the decay energy of the radioisotope for subsequent transfer to the solute. Changes to the solvent, such as dilution with other material, may have a marked effect on the efficiency with which the solvent fulfils these roles.

Solvents fall broadly into three categories:

(i) Effective solvents, e.g. the aromatic hydrocarbons, of which toluene and xylene are by far the most widely used.

(ii) Moderate solvents, e.g. many non-aromatic hydrocarbons. With appropriate scintillator these may result in counting efficiencies of 15–40% of that of toluene.

(iii) Poor solvents; unfortunately this is virtually everything else including most common laboratory solvents such as alcohols, ketones, esters and chlorinated hydrocarbons. Poor solvents usually give solutions with counting efficiencies of $< 1\%$ of that of toluene.

The effect of adding a moderate or poor solvent to a liquid scintillator solution based on an effective solvent is shown in *Figure 5.3*. The effect on the counting efficiency is interesting because most of the samples that one has occasion to count tend to come from the poor solvent category.

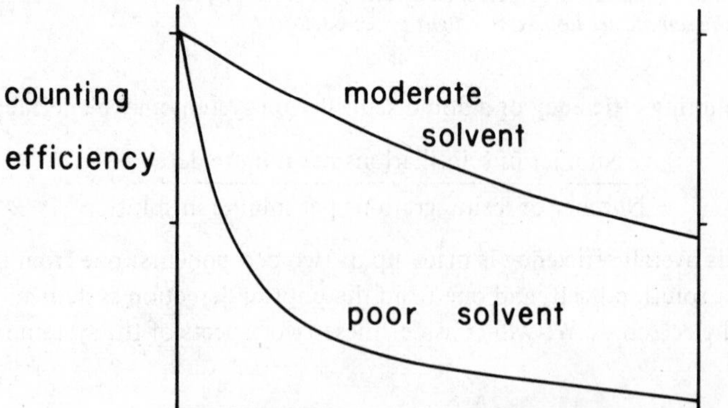

Figure 5.3 Typical reduction of counting efficiency of a liquid scintillator solution as a poor or moderate solvent is added to the solution. The quantitative effect will depend on the nature of the radionuclide present and the scintillator solute in the solution

(2) The primary solute

This is the name given to the scintillator material in the solution. The solute acts as a trap for the radioactive decay energy initially converted into electronic excitation energy by the solvent molecules. The earliest solutes were naphthalene, anthracene and *p*-terphenyl, and large conjugated systems are still the most widely used, and indeed the most efficient. Modern scintillator solutes often have lengthy chemical names and are most frequently known by simple abbreviations. Among the commonest are:

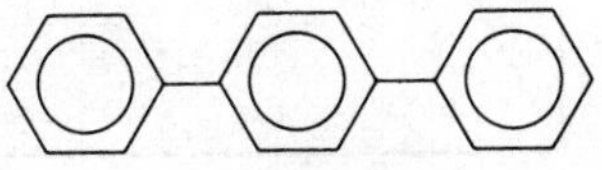

p – terphenyl

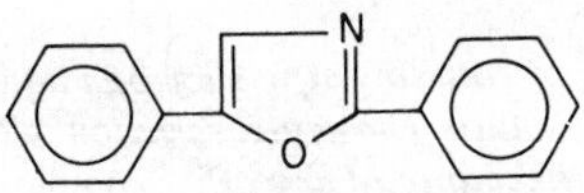

PPO
(1–phenyl, 4–phenyloxazole)

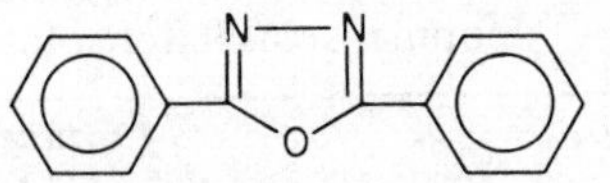

PPD (1–phenyl, 4–phenyl oxadiazole)

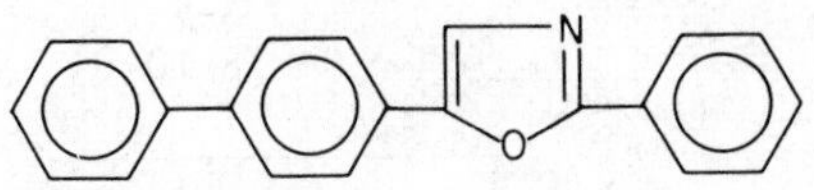

PBO (2–phenyl, 5(4–biphenyl) oxazole)

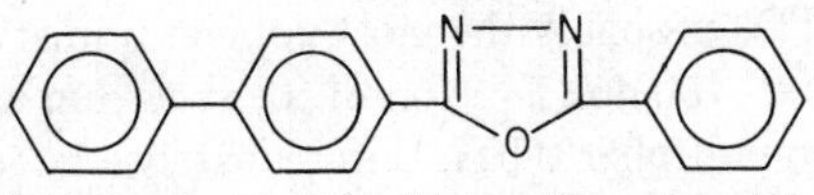

BPD (2-phenyl, 5(4–biphenyl) oxadiozole)

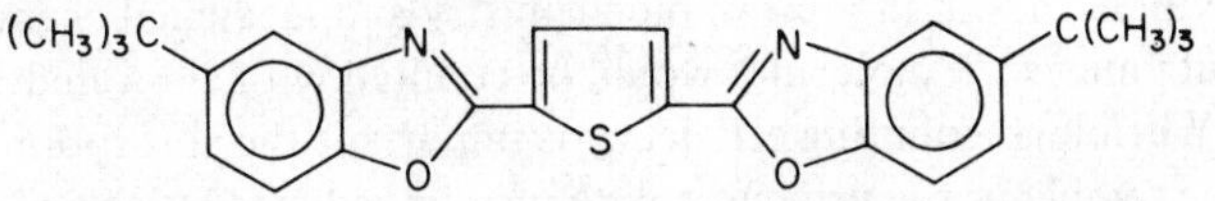

BBOT (2, 5 – di – (5 – t – butyl – 2 – benzooxazoyl) thiophene)

As the concentration of primary solute in a scintillator solution increases so the counting efficiency for a sample of, say, ^{14}C initially increases, as shown in *Figure 5.4*. However, a scintillator solute, being an efficient photon emitter, is also a good photon absorber, so that a concentration is reached at which the number of photons escaping from the solution begins to fall. As a result there exists an optimum solute concentration at which maximum counting efficiency occurs. *Table 5.1* shows primary solute concentrations generally used in popular scintillator solutions.

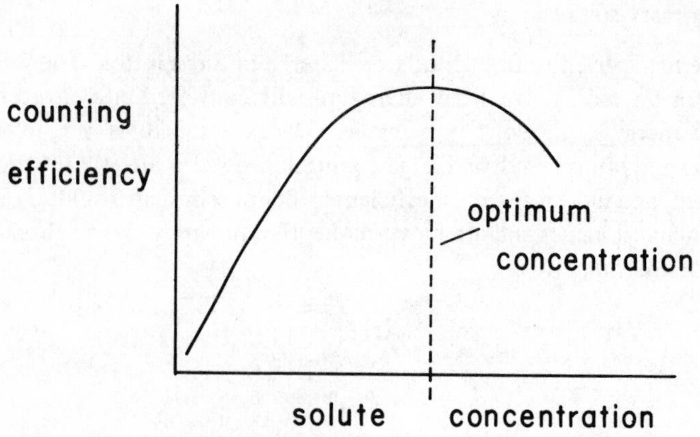

Figure 5.4 Typical variation of counting efficiency of a liquid scintillator solution with the concentration of the primary solute. The actual variation will depend on the nature of the radionuclide present and on the nature of any other compounds in the solution

Table 5.1 Primary solute concentrations of popular scintillator solutions

Solute	*Concentration* ($g\,l^{-1}$)
p-terphenyl	5
PPO	3–7
PBD	8–10
BBOT	7

(3) The secondary solute

The photons emitted by PPO–probably the most popular primary solute–have wavelengths in the range 300–400 nm, a region of the spectrum which is ideal for detection by modern photomultiplier tubes. Unfortunately a large number of molecules have inconvenient photon absorptions in this region of the spectrum, particularly molecules of interest in biochemistry and medicinal chemistry, and samples containing such molecules would be counted with a lowered counting efficiency. When high counting efficiency is important the absorption of photons by either the sample or the primary solute may be reduced by using a secondary solute, which traps the excitation energy from the primary solute and emits photons of a longer wavelength.

The most important secondary solutes are POPOP and its dimethyl derivative. Dimethyl POPOP is 1,4-di-2-(4 methyl, 5 phenyl oxazoyl) benzene

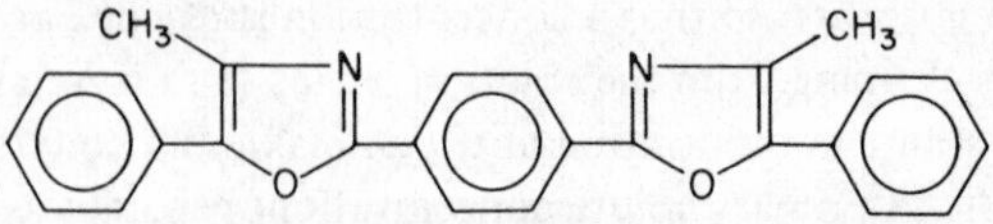

and emits photons in the 400–500 nm range. Both materials have rather low solubilities in common scintillator solvents and are normally used in concentrations of 0.05–0.2 $g\,l^{-1}$.

The combination of solvent and solutes, together with additives which are sometimes used to improve miscibility of the sample material with the solvent, is referred to as a scintillator cocktail. Although an excellent range of scintillator cocktails is available from commercial suppliers, some laboratories find it worthwhile to mix their own and a selection of commonly used recipies is given in *Table 5.2.* Note that purified materials, including the solvent, should be used to make these cocktails if problems such as chemiluminescent photon emission are to be avoided. 'Scintillation grade' solvents and solutes are available from many suppliers, and the slightly higher cost than the laboratory reagent grade is usually worthwhile.

Table 5.2 Typical scintillator cocktails (concentrations in g l^{-1})

Solvent	*Primary solute*	*Secondary solute*	*Additives*	*Sample type*
Toluene	PPO (4–6)	Me_2 POPOP (0.05–0.2)	–	organic soluble
Toluene	butyl PBD (8–12)	– –	– –	organic soluble
p-Xylene	butyl PBD (8–2)	–	Ethanol (100)	aqueous
1,4 Dioxan	PPO (4–6)	Me_2 POPOP (0.2)	Methanol (100) Ethylene glycol (20) Naphthalene (60)	aqueous

5.4 Sample preparation

There are many methods of preparing a sample for liquid scintillation counting. In each case the aim is to achieve a mix between sample and scintillator which will ensure first that the decay radiation deposits its energy within the scintillator solution, and second that the photons emitted during scintillation can escape from the sample bottle and be detected by the photomultiplier tubes. A few of the available methods are described briefly below.

(1) Direct solubilisation

If the radioactive sample happens to be soluble in the scintillator solvent (for example, toluene for organic soluble samples or dioxan for aqueous samples) then a small amount of sample can be dissolved directly in the scintillator cocktail. While this is the simplest method of sample preparation there can be difficulties. For example, when aqueous samples are mixed with a cocktail one occasionally observes the precipitation of the primary solute, a problem which may be exacerbated by refrigeration of the mixture. Many samples can be rendered soluble by a relatively straightforward chemical process, such as complexing with alkyl phosphoric acids, a procedure which has been used for holding radioisotopes of metal ions in scintillator solutions.

(2) Action of a solubiliser

Tissues, proteins, nucleic acids and a wide range of macromolecules may often be converted into a soluble form using a solubiliser. Quaternary ammonium salts are useful solubilisers, although it should be noted that chemiluminescent processes can occur unless highly purified material is used. Some companies supply a 'scintillation grade' range of these salts. Other commercial solubilisers are marketed under names such as Hyamine Hydroxide, Soluene, Protosol and Tissue Solubiliser. Some liquid scintillation cocktails are supplied with a solubiliser already added so that macromolecular materials can be dissolved directly.

(3) Gel counting

While solid insoluble samples may be counted on filter papers carefully positioned within the sample bottle, the result is usually poor in efficiency and reproducibility. It is generally better to count the sample as a fine powder suspended in the scintillator cocktail. Many powders will remain in suspension if broken up using ultrasonics. For dense particles the viscosity of the cocktail may be increased to prevent the settling out of the sample before counting is completed. Thickening agents available for this purpose include aluminium stearate, toluene di-isocyanate* and several polyolefinic resins. In some cases the mixture must be heated to induce thickening. α and low energy β^- emitting isotopes may be difficult to count reproducibly by this method unless the particle size can be controlled, as some self-absorption of the radiation may occur. For the same reason determination of the absolute counting efficiency may be difficult.

(4) Emulsion counting

Liquid samples (usually aqueous solutions) which are not miscible with an aromatic based scintillator cocktail can be dispersed to form an emulsion. As with solid particles in gel counting, the size of the micelles is very important in determining the counting efficiency for low energy emitters, and again ultrasonics may be used to obtain small micelles. Emulsions of this type are generally stabilised by using an emulsifier–typically a polyethoxylated surfactant. Commercially available emulsifiers such as Triton N.104 are excellent, although there have been reports that much cheaper industrial detergents can be just as good. Several ready-made emulsion liquid scintillator cocktails are commercially available, and some can accept quite large quantities of sample. For example, Nuclear Enterprises Micellor scintillator, NE262, can accept up to 40% water. Many of these cocktails can be used for counting a wide range of water-soluble proteins, nucleotides, salts and sugars.

Most samples for liquid scintillation counting are counted in small vials which hold 15-20 ml, although in recent years liquid scintillation counters have become available for counting minivials holding about 5 ml of liquid. Sample vials are usually made of polythene or glass. Polythene vials are disposable and cheap, but cannot be used for holding some organic solvents for long periods. Although they

*Carcinogen

may not look transparent, in fact they transmit the scintillation photons slightly more efficiently than clear glass. Glass bottles allow easier inspection of the sample and can be used to store samples for moderate periods. Ordinary glass bottles can be troublesome for very low activity counting because of the presence of the radioisotope ^{40}K in the glass. Low potassium vials are available from many suppliers, although they are more expensive than ordinary glass bottles. Some laboratories re-use glass sample vials after washing and checking for contamination.

Samples for liquid scintillation counting must be moved around the laboratory (from workbench to counter, and so on) and I prefer to use polythene sample bottles where possible, simply because they bounce when dropped.

5.5 Counting channels

Liquid scintillation counting is most frequently used for counting β^- emitting isotopes and it has been found that the number of photons emitted per β^- decay (and hence the pulse height) is proportional to the energy of the β^- particle. Thus with a suitable photon detection system a liquid scintillation counter operates as a β^- energy spectrometer. The average energies of β^- particles emitted by ^3H, ^{14}C and ^{32}P (three of the most commonly used radioisotopes) are ~ 5 keV, 50 keV and 500 keV respectively. To obtain countable electronic pulses from photomultiplier tubes operating under fixed conditions, and viewing scintillations which vary in intensity by more than three orders of magnitude, it is common practice to follow the photomultiplier tubes with either a logarithmic amplifier or several separated linear amplifiers in parallel (each with a gain optimised for one particular range of β^- energies).

With a logarithmic amplifier the amplified pulse heights produced by the β^- decays of the three radioisotopes have distributions as shown in *Figure 5.5*. Using

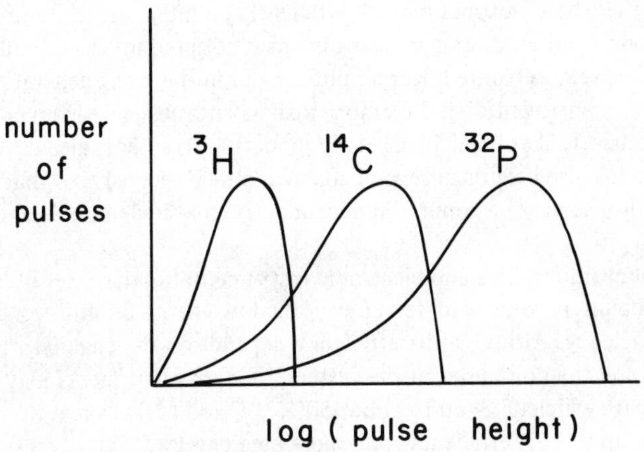

Figure 5.5 Pulse height distributions obtained during decay of ^3H, ^{14}C *and* ^{32}P *in a liquid scintillation counter*

an upper and lower level pulse height discriminator it is easy to select a range of pulse heights (equivalent to a range of β^- decay energies) for counting, so that pulses from one radionuclide are counted with high efficiency while pulses from noise or other isotopes are essentially ignored. For example, with the discriminators set as shown in *Figure 5.6* the ^{14}C isotope is counted while decays of ^{3}H or ^{32}P are largely ignored. The range of pulse height accepted by the counter is that which lies in the window or channel between the two discriminators. Clearly the channel shown in *Figure 5.6* would allow ^{14}C to be counted in the presence of ^{3}H or ^{32}P without major interference from the latter isotopes.

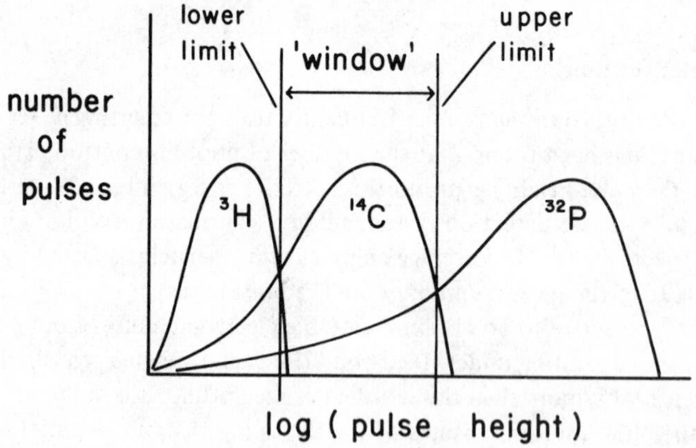

Figure 5.6 Positioning of pulse height limits to set up a window or channel suitable for counting ^{14}C *decays without major interference from* ^{3}H *or* ^{32}P

Most commercial liquid scintillation counters have facilities for switching to lower and upper level discriminators which have been preset as ^{3}H, ^{14}C or ^{32}P channels, so that these isotopes may be efficiently counted either alone or in the presence of one another. A similar arrangement is adopted in liquid scintillation counters with several separate linear amplifiers and in the latest generation of microprocessor controlled/digital memory instruments available from companies such as Beckman, Packard and LKB. Many modern instruments also have the ability to count pulse heights in several channels simultaneously, so that two or more radioisotopes may be counted independently in a single sample and at the same time.

Modern photomultipliers and electronic instrumentation have resulted in liquid scintillation counters which can detect even the low energy β^- emissions of ^{3}H with high efficiency. Although the efficiency depends on the channel width used for counting and the noise level of the system, modern instruments may count ^{3}H samples with efficiencies up to about 50%. ^{14}C and ^{35}S samples may be counted with up to 95% efficiency, and most high energy β^- emitters (e.g. ^{32}P) with up to 99% efficiency. In practice the counting efficiencies are often lower than these values because of the presence of impurities, sample solvents and other materials from the 'poor solvent' category. This reduction of counting efficiency is called quenching.

Quenching reduces the number of photons emitted from a scintillator solution for a given radioactive decay energy. The magnitude of the electrical pulse reported by the photomultiplier tubes is reduced as a result, so the efficiency of counting for a particular radioisotope varies with the degree of quenching within the solution. Quenching effects in fact fall into two categories: (i) chemical quenching, caused by de-excitation of electronically excited molecules which would otherwise give rise to emitted photons; and (ii) colour quenching, caused by absorption of emitted photons by materials in the solution—usually a coloured sample. While (ii) may sometimes be overcome by the use of a secondary solute such as POPOP, which moves the wavelength of the emitted photons to a different region of the spectrum, there is usually little that can be done about (i) except to restrict the amount of quenching agent added to the scintillator cocktail.

For the low energy β^- emitting isotopes, particularly ^3H, quenching is often severe. Oxygen gas dissolved in a scintillator cocktail can reduce counting efficiency by 20% for a ^3H sample and for this reason scintillator cocktails are often flushed with nitrogen or argon before use. Samples stored in refrigerators are particularly prone to contamination by water during sample preparation, again causing significant quenching problems. In general there are few occasions on which one can afford to ignore the effect of quenching on the counting efficiency in liquid scintillation counting, and it is usual practice to determine the counting efficiency with which every sample is counted.

5.6 Estimation of counting efficiency

Within the past few years the number of ways of estimating counting efficiency in liquid scintillation counting has increased. Fortunately the methods employed on most commercial instruments are variations on the two basic techniques described below. As there is a sense in which all lowering of counting efficiency below 100% is a result of quenching, the procedure of estimating the efficiency of counting is frequently referred to as the estimation of the quench correction.

(1) The sample channels' ratio method (SCR)

Because all liquid scintillation counters have facilities for setting channels for particular isotopes, any individual isotope may be counted in two separate channels as illustrated in *Figure 5.7* for an unquenched ^{14}C sample. In this example channel A is a standard ^{14}C channel and the counts recorded in this channel are counts required for the sample activity measurement. A second channel, B, can be set to record about one-tenth of the β decays counted in channel A.

When quenching occurs the magnitude of every pulse is affected, so that the whole pulse height spectrum produced by the β^- decays moves to lower pulse heights, as shown in *Figure 5.8*. Clearly the ratio of counts recorded in the two

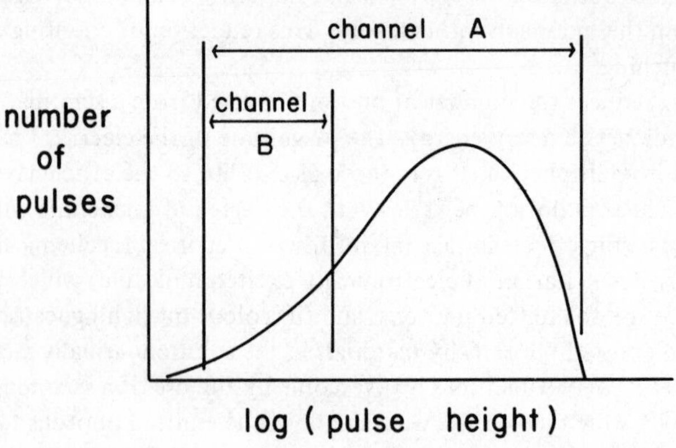

Figure 5.7 Two channels set up so that ^{14}C *may be counted in channel* A *and a sample channels ratio may be obtained using* A/B

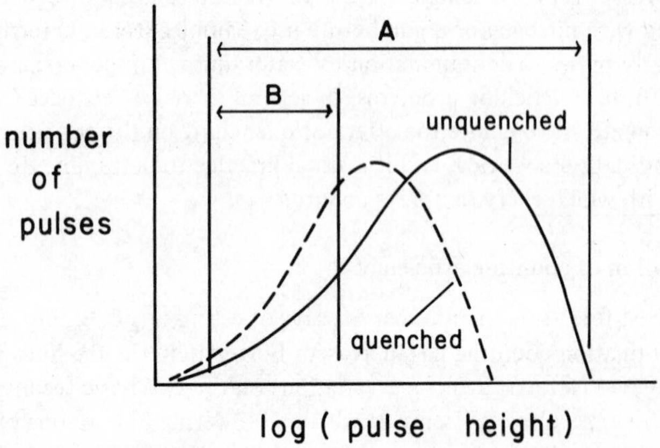

Figure 5.8 Effects of quenching on the position of the pulse height distribution. Note that quenching results in a greater portion of the distribution being counted in channel B, *thus lowering the ratio* A/B

channels, A/B, is different in the quenched solution from that in the unquenched solution (~ 10 in the unquenched spectrum and ~ 3 in the quenched spectrum of *Figure 5.8*). In practice a calibration graph of absolute counting efficiency in channel A versus the channels ratio, A/B, is drawn using standards of known activity and varying degrees of quenching. A typical 'quench correction curve' of this type is shown in *Figure 5.9*. Once the calibration curve has been drawn, subsequent samples are simply counted in the two channels A and B, the channels ratio calculated and the counting efficiency in channel A read directly from the

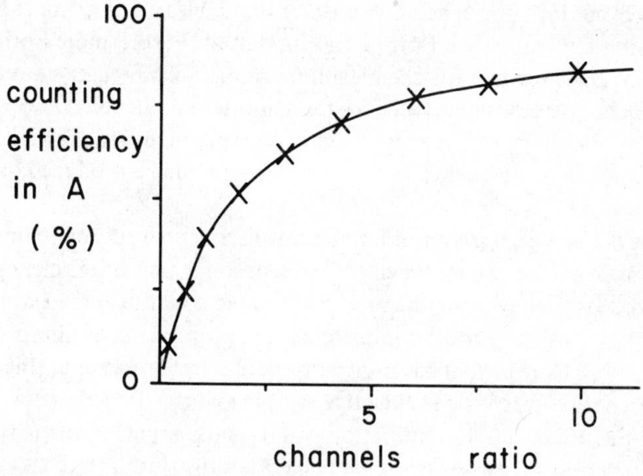

Figure 5.9 Typical calibration curve for efficiency estimation, often called a 'quench correction curve'

calibration graph.

Most liquid scintillation counter suppliers will supply series of quench samples of radioactive material in scintillator cocktail (usually 10 or 12 samples of identical activity but differing degrees of quenching) so that the quench correction curves—there must clearly be a different curve for each isotope and each counting channel—may be checked or redrawn at regular intervals.

The disadvantage of the sample channels ratio method of efficiency estimation is that for low activity samples the time required to record a statistically useful number of counts in the smaller channel (channel B in *Figure 5.8*) may be long. This problem is largely overcome by the second method.

(2) The external standard channels' ratio method (ESCR)

The ESCR method is based on the same reasoning as the SCR method. The difference between the two is that for the ESCR method the pulse spectrum used for the ratio measurement—not, of course, for the sample activity measurement—is provided by electrons scattered within the scintillator solution by an external γ-ray source (usually ^{137}Cs or ^{226}Ra). Clearly this approach requires three separate measurements: one of the sample in the appropriate isotope channel; one of the sample in the two preset ESCR channels, and one of the counts produced by the γ source in the two ESCR channels, the second set of counts being subtracted from the third before the ratio is calculated. Both sets of counts in the ESCR channels may be performed quite rapidly since a high activity γ source (typically several hundred kBq) may be used.

As before, a calibration curve of counting efficiency versus channels ratio is drawn using quenched standards of known activity, and this is used to determine the absolute counting efficiency of samples on a routine basis.

Whichever method of quench correction is used, clearly there must be a different calibration curve for each isotope to be counted. Furthermore both methods, while usually quite reliable for chemical quenching, must be used cautiously for colour quenching, because in the latter the scintillation photons may be selectively absorbed and the shape of the pulse height spectrum modified. Certainly for accurate work a separate quench calibration curve should be produced for coloured samples.

Many of the new generation of liquid scintillation counters contain microprocessors and may be programmed to perform some kind of efficiency estimation, in some cases converting the recorded sample count into the estimated absolute activity of the sample complete with appropriate error limits on the reported value. The principal advantage of liquid scintillation counting, i.e. the lack of a barrier between the radioactive sample and the detecting device, has resulted in the use of liquid scintillation counters for counting virtually all α and β^- decay isotopes. Recent developments have led to scintillator cocktails containing heavy-metal compounds, which can be used for counting γ-emitting nuclides–particularly low energy γ-emitters such as ^{125}I. In fact it is probably true to say that the single most useful item of counting equipment for any radiochemical laboratory is a liquid scintillation counter, as there are very few radionuclides which cannot be counted with an acceptable efficiency in such an instrument.

5.7 Cerenkov counting

While virtually any radionuclide can be counted using liquid scintillation counting, problems can arise in the sample preparation stage simply because an organic scintillator material needs to be in intimate contact with the radioactive sample. While the use of gel or emulsion counting may often overcome problems of incompatibility, there is an alternative and very useful technique for counting energetic β^+ and β^- decay isotopes which does not involve the use of a scintillator material at all. The method is based on the Cerenkov effect, which is the emission of a bluish-white light (Cerenkov radiation) when an electron or positron travels through a medium with a velocity which is greater than the velocity of light in that medium. (The velocity of light in a vacuum, c, cannot be exceeded: light travels more slowly through matter.)

The minimum particle velocity, and hence the threshold energy, at which the Cerenkov effect is observed depends on the refractive index of the medium involved. In most common solvents the Cerenkov threshold for β^- particles lies between 0.15 and 0.27 MeV. In water, for example, the minimum β^- particle energy which results in the emission of Cerenkov radiation is 0.263 MeV. As energetic β^- particles pass through matter emitting Cerenkov radiation they lose energy, primarily by ionisation and electronic excitation processes, and quickly fall below the Cerenkov threshold. Consequently the Cerenkov radiation emitted by β^- particles in a solution of a radioisotope appear as brief flashes of light, which may be detected and counted in much the same way as the scintillations in liquid scintillation counting.

For energetic β^- emitters such as ^{24}Na and ^{32}P, most of the β^- particles emitted are above the Cerenkov threshold for water (84% and 90% respectively), so that these isotopes may be detected and counted as aqueous solutions simply by placing a vial of the aqueous solution in a liquid scintillation counter. As no scintillator cocktail is required, larger sample volumes can be accommodated in the standard vial than would be possible in conventional liquid scintillation counting.

Unless one does a large amount of Cerenkov counting it is usually most convenient to count Cerenkov radiation in the ^3H channel of a liquid scintillation counter—this being the channel covering the lowest pulse height region. Of course, the counting efficiencies for different isotopes in this channel must be determined experimentally, but a rough guide to the efficiencies to be expected in water are given in *Table 5.3*. It should be noted that the manner in which Cerenkov radiation is emitted results in the efficiency of detection in a liquid scintillation counter being a function of sample volume, the optimum volume in many instruments being about one half the volume of a standard liquid scintillation vial. Hence all samples to be counted by Cerenkov counting should be of the same volume, additional solvent being added to some samples if necessary. It is clear from *Table 5.3* that counting efficiencies attainable by Cerenkov counting are not as high as would be expected using liquid scintillation counting. Neverthless, the great simplicity of sample preparation may often go some way to compensating for this.

Table 5.3 Typical Cerenkov counting efficiencies for various radionuclides in ^3H channel of liquid scintillation counters

Radionuclide	*Cerenkov counting efficiency** (%)
^{137}Cs	2
^{36}Cl	2
^{47}Ca	7
^{40}K	14
^{24}Na	18
^{32}P	25

*Samples in aqueous solution.

Solvents other than water are also used for Cerenkov counting. Saturated hydrocarbons, halocarbons and alcohols are frequently useful, but if unsaturated or aromatic solvents are to be involved one must take particular care to check the counting efficiency, as a weak fluorescence from the solvent may boost the counting efficiency dramatically. While chemical quenching does not occur in Cerenkov counting, colour quenching can be a problem for coloured samples. Bleaching can often solve the problem, although where this is not practical the colour quenching can be accounted for using techniques which are analogous to the channels' ratio methods of liquid scintillation counting.

As the counting efficiency is largely unaffected by the chemical content of the sample solution, Cerenkov counting is often used for counting materials that hav been digested by rather drastic chemical means. For example, labelled biological material may have been treated with perchloric acid before counting. Note that low level chemiluminescence is a fairly widespread phenomenon and may produce misleading count rates from systems which are not obviously emitting light. Control samples, from which the radioisotopes have been omitted, provide a useful check for such problems.

Bibliography

Turner, J. C., *Sample preparation for Liquid Scintillation Counting*, Review 6, The Radiochemical Centre, Amersham (1971)

Birks, J. B., *An Introduction to Liquid Scintillation Counting*, and *Solutes and Solvents for Liquid Scintillation Counting*, (combined edn), Koch-Light Laboratories Ltd, Colnbrook (1975)

Birks, J. B., *The Theory and Practice of Scintillation Counting*, Pergamon, Oxford (1964)

Horrocks, D. L., *Applications of Liquid Scintillation Counting*, Academic, New York (1974)

Long, E. C., *Liquid Scintillation Counting, Theory and Techniques*, Beckman Instruments, High Wycombe (1976)

6 γ Spectrometry

6.1 γ emissions

The energies of γ photons emitted by excited nuclei are characteristic of those nuclei, just as the energies of visible and UV (ultraviolet) photons emitted by excited atoms are characteristic of those atoms. For atoms the sophisticated science of atomic spectroscopy has been developed to the point where elemental analysis of materials may be undertaken directly from spectra. In much the same way the techniques of γ-ray spectrometry have evolved so that a large number of γ-emitting radionuclides can be identified from their γ-photon spectra. However, before describing the details of γ spectrometry we must consider briefly the variety of ways in which γ-emitting nuclides are formed.

Electronically excited atoms and molecules are most frequently generated from ground state species following the absorption of UV or visible light photons. Excited nuclei can be produced in an analogous manner, provided a source of γ radiation of the correct energy is available. Sources of γ photons with a wide range of energies are scarce, so that for practical purposes the production of excited state nuclei by direct excitation from the ground state is of little interest.

γ-emitting nuclides are most often formed by nuclear reactions. Quite apart from the (n, γ) class of reactions, which initially form the very short-lived excited state nuclei responsible for the γ emission, many nuclear reactions result in the formation of relatively long-lived excited states–isomeric or metastable states–which may decay either wholly by γ-emission, or by competing processes of γ emission and α or β decay. A typical example is the formation of ^{34m}Cl by fast neutron reaction with ^{35}Cl, i.e.

$$^{35}\text{Cl}\,(\text{n}^*, 2\text{n})\ ^{34m}\text{Cl}$$

The product nucleus has a half-life of ~ 30 minutes and decays partly by β^+ emission and partly by γ photon emission. An energy level diagram for ^{34m}Cl decay is shown in *Figure 6.1*, where the percentage of all decays which follow the different decay routes is indicated. Many γ-emitting radionuclides are deliberately generated by nuclear reactions with stable nuclides, so that the radionuclides may be identified from the γ-photon energies, and the identities of the stable precursor

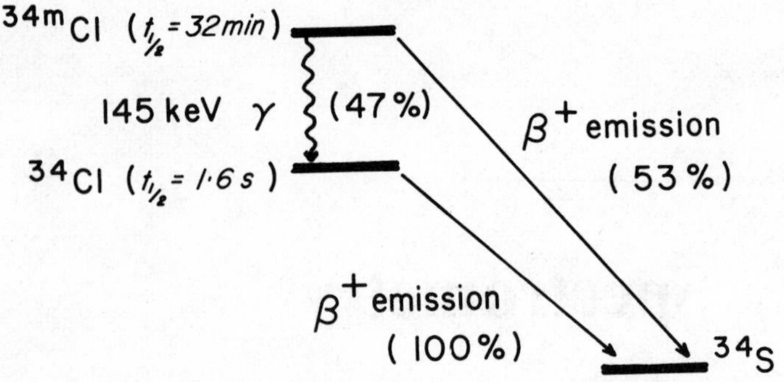

Figure 6.1 Partial decay scheme for ^{34m}Cl

nuclei inferred from those of the γ-emitting products. This general procedure is called activation analysis, and is widely used for elemental analysis as it can often be applied without damaging the sample.

α and β decay of radioisotopes provides another source of γ-emitting nuclides. The decay products of α and β decays are frequently formed in excited states, so that most α and β decays are accompanied by the emission of γ photons from excited daughter nuclei. For example, the β^- decay of ^{137}Cs results in the formation of some excited state daughter, ^{137m}Ba, which in turn emits a 662 keV γ photon as it relaxes to the ground state. The energy levels of the relevant nuclei are illustrated in *Figure 6.2.*

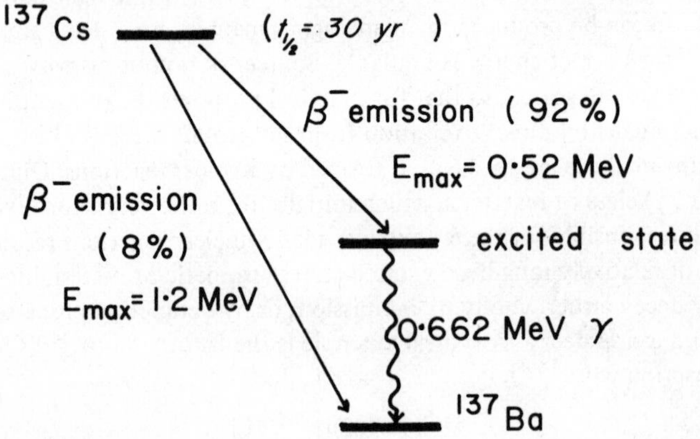

Figure 6.2 Partial decay scheme for ^{137}Cs

6.2 The γ spectrometer

A γ spectrometer contains the elements shown in *Figure 6.3*. The γ photon detector may be any device which converts the γ-photon energy into an electrical pulse, and is usually a γ-scintillation detector (e.g. NaI (T1) crystal and photomultiplier) or a semiconductor detector (e.g. a Ge–Li detector). The essential

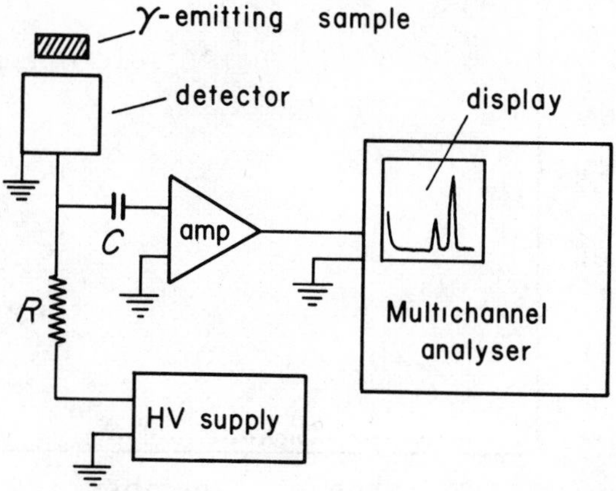

Figure 6.3 Schematic arrangement of basic circuit elements of a γ-photon spectrometer

feature of the detecting device is that it should produce an electrical pulse whose magnitude is directly proportional to the energy of the γ photon which initiates the pulse. When either of the above detectors is used the magnitude of the electrical pulse is amplified by a simple pulse amplifier before further treatment. Up to this point the system has been the same as the apparatus described in chapter 3 for the counting of γ-emitting isotopes. Essentially the spectrum of energies of γ photons entering the detector has been converted into a spectrum of electrical pulses leaving the pulse amplifier. It is this electrical pulse spectrum which is recorded and analysed in γ spectrometry.

There are several ways of recording the electrical pulse spectrum. Probably the most commonly used technique involves a multichannel analyser (MCA). This is an electronic device which contains several hundred pulse counters. Each pulse which enters the MCA input is analysed by a pulse height analyser and then sent to one of the pulse counters, each counter being used to count pulses within a different range of pulse heights. After several thousand pulses have been analysed some counters will have recorded many counts, while others may have recorded very few. Those containing many counts will be those whose pulse height range corresponds to the photon energies of γ radiation which has entered the detector.

Most multichannel analysers display the contents of their counters, or channels, on a screen in the form of a graph of the number of counts recorded in a channel against the mean pulse height counted in that channel. Close examination of such a display will usually reveal that the display is made up of several hundred dots on the screen, one dot for each channel in the MCA. A typical (idealised) MCA display following γ photon detection with a NaI (T1) scintillation detector is shown in *Figure 6.4*. In view of the proportionality between the γ-photon energy and the pulse height of the amplified electrical pulses, *Figure 6.4* may also be regarded as an approximation to the spectrum of γ-photon energies which gave rise to that display. The actual (but again idealised) γ-photon spectrum would more correctly be that shown in *Figure 6.5*.

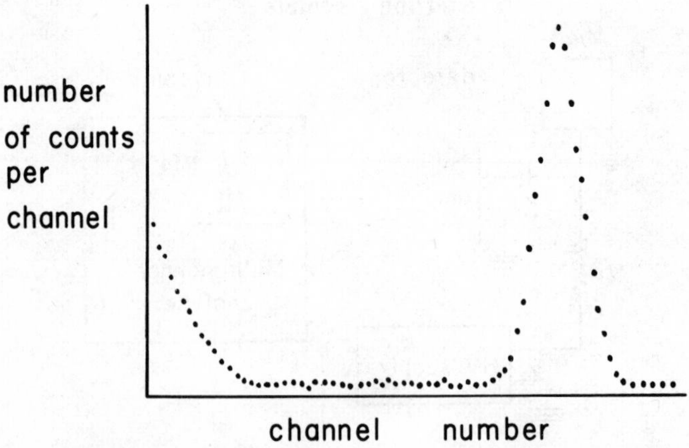

Figure 6.4 Typical display of pulse height distribution spectrum on a MCA. The number of counts recorded in each pulse height range (channel) is shown as a dot on a display screen. Distribution shown is a simplified form of what may be anticipated using a NaI(T1) *crystal detector*

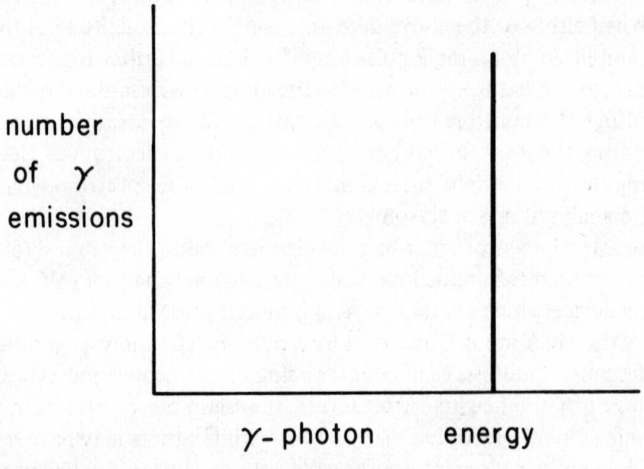

Figure 6.5 The actual energy spectrum of γ photons emitted by the sample which gave the pulse height distribution shown in Figure 6.4

The pulse height spectrum obtained using a NaI (T1) crystal scintillation detector exhibits peaks which are clearly much broader than the spread of energies in the actual γ photon spectrum. This is because the resolution of a crystal scintillation detector, defined as the ratio of the width of the peak (in mV) to its centre position (in mV), is usually rather poor, commonly ~8–10%. Much better resolutions are possible with semiconductor detectors, where resolutions of < 1% can be achieved. The pulse height spectrum shown in *Figure 6.6* represents an idealised MCA display obtained from the γ-photon spectrum shown in *Figure 6.5* using a

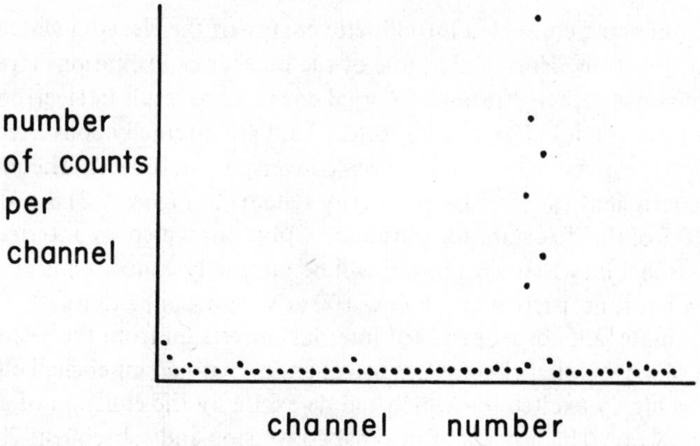

Figure 6.6 Display of pulse height distribution which may be anticipated using a semiconductor detector to analyse the γ-photon spectrum of Figure 6.5. Note the much narrower peak width in this case compared with that in Figure 6.4

Ge–Li semiconductor detector. In spite of the obvious improvement in the peak width obtained with semiconductor detectors, the lower detection efficiency for γ photons with energies above ~100 keV has resulted in the continued use of crystal scintillation detectors for the bulk of γ spectrometry associated with radiochemistry.

Most commercial MCA instruments have counters in multiples of 100 (e.g. 200 or 400 channel analysers) or 128 (e.g. 512 or 1024 channel analysers). The ability to display the spectrum on a screen is usually supplemented by a facility enabling a print out of the count recorded in each channel to be obtained. Some modern MCAs contain a microprocessor and have options such as automatic calibration of channel number in terms of γ-photon energy, so that peaks in a γ spectrum can be directly identified with particular energies. In less sophisticated instruments the user must calibrate the MCA channels against energy, normally by recording γ spectra of standard isotopes with well known γ energies. Suppliers of radioisotopes, such as the Radiochemical Centre, market small encapsulated samples of nuclides for this purpose.

6.3 γ spectra

The peaks recorded by a γ spectrometer and which correspond to γ photons emitted by a radionuclide are called photopeaks. Comparatively few γ spectra are found to consist of a straightforward series of photopeaks, so we must consider the origins of the other peaks and bumps which are characteristic of most practical γ spectrometry.

The first such complication has its origin in the loss of γ photons by a process known as internal conversion. This occurs when a γ photon leaving the nucleus undergoes a photoelectric collision with an extranuclear electron, the energy of the γ photon being converted into kinetic energy of the electron which is then ejected from the atom. Thus while some of the nuclear de-excitations give rise to γ-photon emissions, others, through internal conversion, result in electron emission. The ratio of the number of γ photons which are internally converted to the number which are not is called the internal conversion coefficient. The internal conversion coefficient for ^{137m}Ba (see decay scheme in *Figure 6.2*) is ~0.11, so that about 10% of the de-excitations produce γ photons which are internally converted. The probability that a γ photon will be internally converted is greatest for low-energy γ photons, particularly below 100 keV, and can be unity.

The most important consequence of internal conversion from the γ-spectrometry point of view is that the process leaves an ion with an inner shell electron vacancy, i.e. a highly excited ion which can de-excite by the emission of a characteristic X ray. The process of internal conversion and subsequent X-ray emission is illustrated in *Figure 6.7*, and the similarity between this process and X-ray emission following electron capture decay (*Figure 1.2*) may be noted. A typical γ spectrum containing a characteristic X-ray photopeak is shown in *Figure 6.8*.

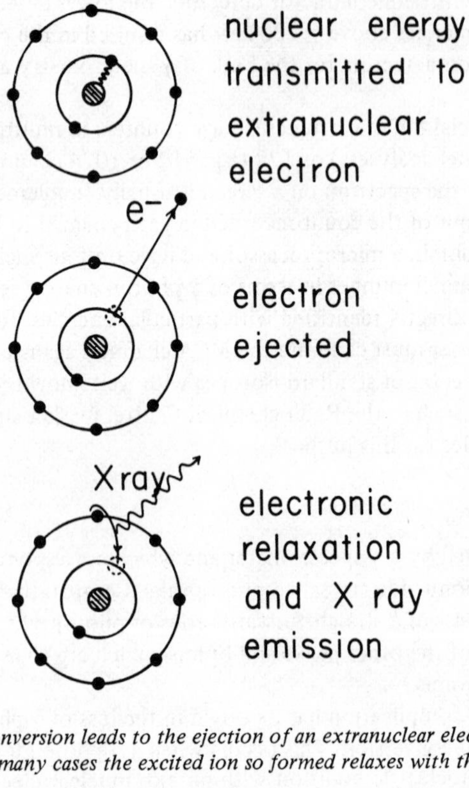

Figure 6.7 Internal conversion leads to the ejection of an extranuclear electron rather than γ-photon emission. In many cases the excited ion so formed relaxes with the emission of an X-ray photon

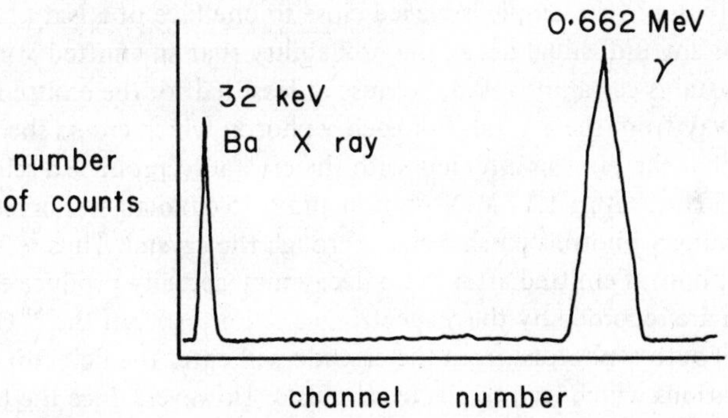

Figure 6.8 Idealised γ spectrum of ^{137}Cs *sample expected using a crystal scintillation detector. Spectrum shows* Ba *X-ray peak which arises following internal conversion of* ~ 10% *of the* ^{137m}Ba *decays*

6.4 Photon summing

A second complication arises in γ spectrometry because many nuclides produced in excited states are produced in more than one excited state, or in a state which is higher in energy than the first excited state. For example, ^{60}Co decays by β^- decay to produce an excited state of ^{60}Ni. As this excited state relaxes to the ground state of ^{60}Ni, γ photons of 1.17 and 1.33 MeV are emitted. A decay scheme showing the energy levels involved is shown in *Figure 6.9*. More than 99% of the

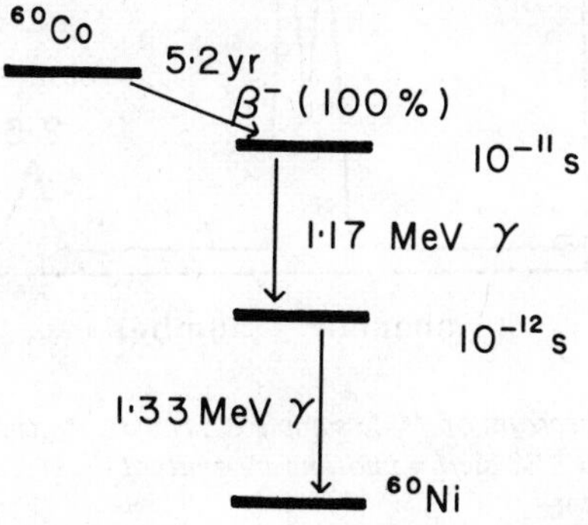

Figure 6.9 Partial decay scheme for ^{60}Co. *The approximate lifetimes of the two excited states of* ^{60}Ni *are included*

^{60}Co decays produce ^{60}Ni in the second excited state, which has an energy of 2.5 MeV greater than the ground state nucleus. This excess energy is lost by emission of a 1.17 MeV γ photon, which leaves the ^{60}Ni nucleus in its first excited state (1.33 MeV above the ground state), followed very rapidly by the emission of a 1.33 MeV γ photon, which leaves the nucleus in its ground state. γ photons emitted in very rapid succession are said to be in cascade.

Suppose that a ^{60}Co sample is placed close to one face of a NaI (T1) crystal detector. For any individual decay the probability that an emitted γ photon will enter the crystal is certainly < 0.5, because at least half of the emitted photons will travel away from the crystal. For each γ photon which enters the crystal the probability that the photon interacts with the crystal to produce a scintillation may be ~ 0.5 (for, say, a 1.17 MeV photon and a 75 cm diameter crystal), many of the remaining γ photons passing clear through the crystal. Thus $\sim 25\%$ of the 1.17 MeV γ photons emitted after ^{60}Co decay may actually produce electrical pulses which are recorded by the γ spectrometer. For $\sim 6\%$ of the ^{60}Co decays (25% x 25%) both γ photons from the cascade will enter the detector and produce scintillations which lead to electrical pulses. However, since the two γ photons are emitted simultaneously (average time between the 1.17 and 1.33 MeV γ emissions after ^{60}Co decay is $\sim 8 \times 10^{-13}$ s), the scintillations produced when both γ photons interact with the crystal cannot be separated, and the resultant electrical pulses are as large as those produced by a 2.50 MeV γ photon.

Thus the γ-photon spectrum recorded for ^{60}Co is found to contain three peaks, at 1.17, 1.33 and 2.50 MeV, as shown in the part spectrum in *Figure 6.10*. The photopeaks at 1.17 and 1.33 MeV are nearly the same height, i.e. about the same

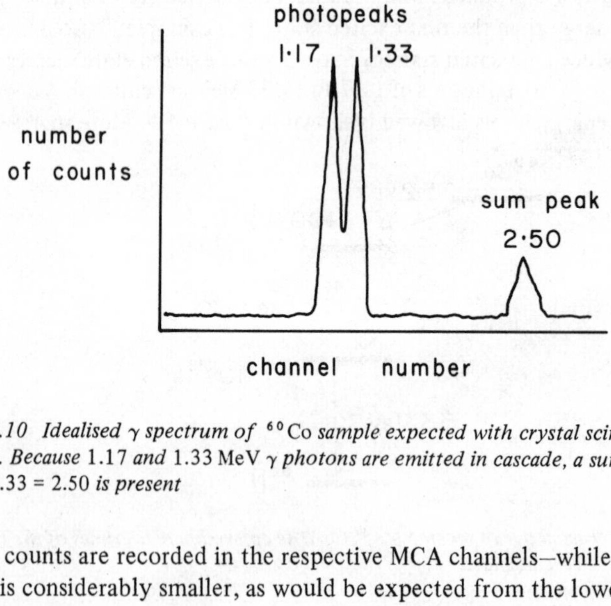

Figure 6.10 Idealised γ spectrum of ^{60}Co sample expected with crystal scintillation detector. Because 1.17 *and* 1.33 MeV *γ photons are emitted in cascade, a sum peak at* 1.17 + 1.33 = 2.50 *is present*

number of counts are recorded in the respective MCA channels–while the 2.50 MeV peak is considerably smaller, as would be expected from the lower probability of both photons producing scintillations simultaneously. The 2.50 MeV peak is called a sum peak, as it is produced by the summing of pulses initiated by two different photons.

Summing can always occur when several γ photons are emitted in cascade over a very short time period, usually < 1 μs. In fact the presence of a peak in a γ spectrum at an energy which corresponds to the sum of the energies of two other peaks is usually indicative of a cascade process. However, if peaks are produced by different processes summing cannot occur (except by accident, in which case the sum peak will be too small for detection). For example, the β^- decay of ^{59}Fe leads to the formation of two excited states of ^{59}Co, as shown in the decay scheme of *Figure 6.11*. Here summing can occur between the 0.19 and 1.10 MeV γ photons to produce a sum peak at 1.29 MeV. However, summing cannot occur between the 0.19 MeV photon and the peak at 1.29 MeV, since the 0.19 MeV photon is an alternative to the 1.29 MeV photon and not an emission in cascade with a 1.29 MeV photon. Part of the spectrum obtained following the decay of ^{59}Fe is shown in *Figure 6.12*.

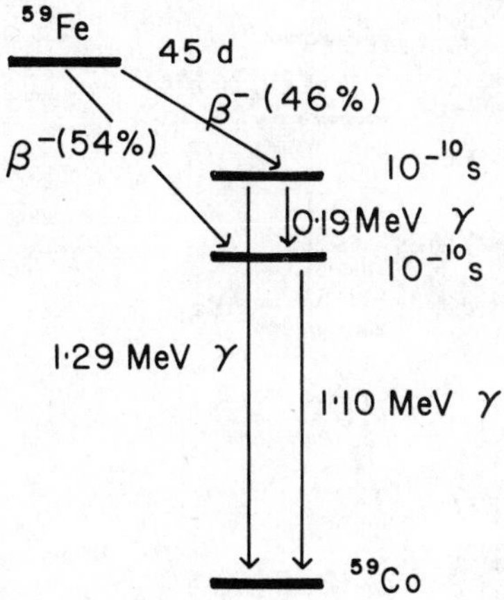

Figure 6.11 Partial decay scheme for ^{59}Fe. *The approximate lifetimes of the two excited states of* ^{59}Co *are included. Note there are two possible γ emissions from the upper excited state*

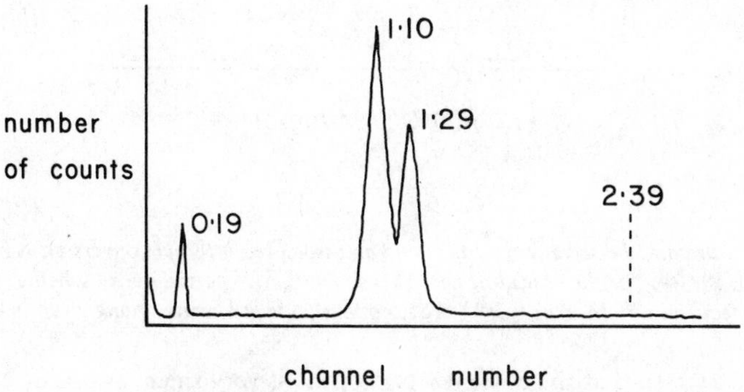

Figure 6.12 Idealised γ spectrum of ^{59}Fe *sample expected with a crystal scintillation detector. The peaks at* 0.19 *and* 1.10 MeV *are γ-photo peaks, whereas that at* 1.29 MeV *has contributions from* 1.29 MeV *γ photons and from summing between the* 0.19 *and* 1.10 MeV *cascade photons. Note the absence of a sum peak at* 2.39 MeV

In addition to summing occurring between cascade γ photons, sum peaks may also arise by summing between γ photons and X rays associated with internal conversion. Because the probability of internal conversion increases as the de-excitation energy decreases, this is a complication most commonly associated with low energy γ processes. For example, ^{132}Te (tellurium) decays to a short lived excited state of ^{132}I, which in turn cascades down to the ground state as shown in *Figure 6.13a*. A γ spectrum of the emissions from this process is shown in *Figure 6.13b*

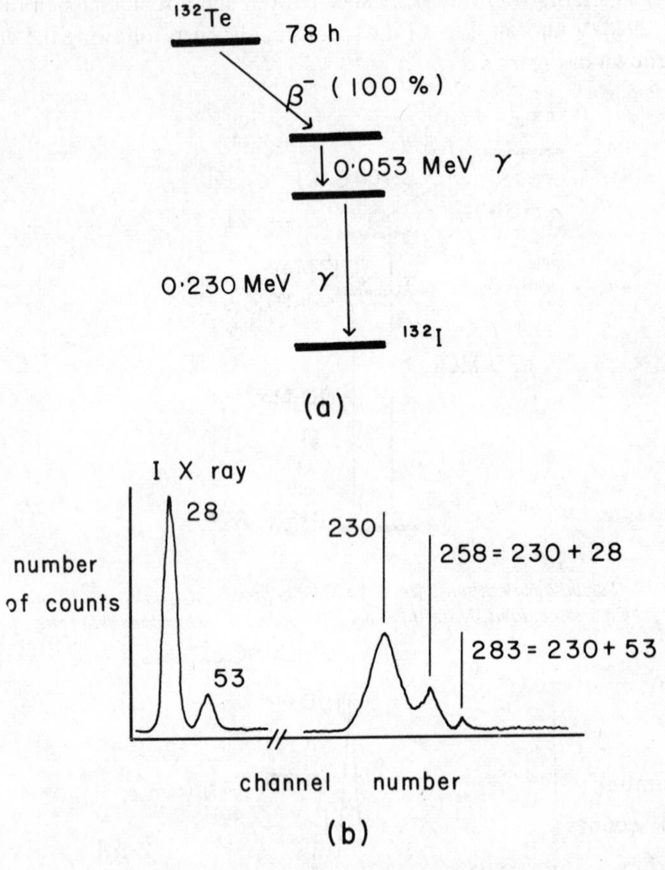

Figure 6.13 Summing between a γ photon and an associated internal conversion X-ray photon demonstrated for the cascading of ^{132}I. *a, The* ^{132}Te *partial decay scheme. b, The idealised γ spectrum obtained from a* ^{132}Te *sample, the peak energies being given in* keV

where it will be seen that the 53 keV γ photopeak is very small, as this de-excitation undergoes extensive internal conversion. The excited iodine ion which remains rapidly emits a characteristic X ray of 28 keV, and in fact this forms the major peak of the spectrum. Now although summing does occur between the 230 keV γ photon and 53 keV γ photon emitted in cascade, the sum peak at 283 keV is considerably smaller than the sum peak at 258 keV formed by summing between the 230 keV γ photon and the 28 keV characteristic X ray.

The increasing use of semiconductor detectors, with their excellent resolution and sensitivity at low energies (10–100 keV), has greatly increased the importance of characteristic X rays and the summing which occurs between these and other γ photons. This is particularly true for element identification by activation analysis.

6.5 Pair production

A third complicating feature of γ spectrometry arises because of the tendency of γ photons with energy in excess of 1.022 MeV to turn spontaneously into electron–positron pairs, a process known as pair production and written:

$$\gamma \xrightarrow{E_\gamma > 1.022\ \text{MeV}} e^+ + e^-$$

Any photon energy in excess of 1.022 MeV appears as kinetic energy of the charged particles. Because pair production is confined to rather high energy γ emitters its consequences are normally observed when large scintillation crystals are used for detection. The main effects arise when the e^+ e^- pair are created within the detector. In this case either all of the original γ-photon energy, E_γ, will be detected as the charged particles lose their kinetic energy and the two 0.511 MeV photons produced by the e^+ annihilation are detected, or one or both of the two 0.511 MeV annihilation photons will escape from the crystal without being detected. Thus the γ spectrum will contain peaks at the original γ-photon energy, E_γ, and escape peaks at $E_\gamma - 0.511$ MeV and $E_\gamma - 1.022$ MeV.

The escape peaks resulting from pair production can be seen in the γ spectrum of ^{24}Na, the decay scheme for which is shown in *Figure 6.14a*. The spectrum recorded with a 7.5 cm NaI (T1) crystal is shown in *Figure 6.14b*. Escape peaks which originate from the 2.76 MeV γ photons are seen at 2.76 – 0.511 = 2.25 MeV and 2.76 – 1.022 = 1.74 MeV. No escape peaks originating from 1.37 MeV γ photons are visible on the scale used, because although pair production probability rises with energy above 1.022 MeV it does not become significant until $E_\gamma > \sim 1.8$ MeV. Similarly while escape peaks could be associated with the 4.13 MeV sum peak, in practice these are normally too small to observe.

Note that pair production and position annihilation may occur outside the crystal detector, for example, in lead shielding or construction material. One of the annihilation photons so produced may enter the detector and give rise to a small peak at 0.511 MeV. This effect can usually be minimised by keeping shielding and other material some distance from the detecting crystal, a procedure which normally rules out the use of most commercially available crystals intended for simple γ-scintillation counting.

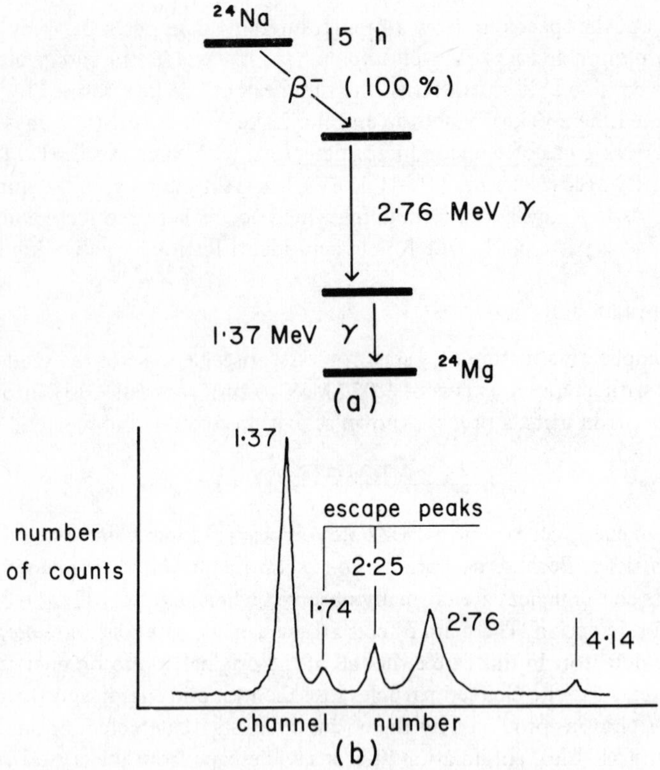

Figure 6.14 Escape peaks resulting from pair production by high energy γ photons. Either or both of the two annihilation photons may escape detection. a, A partial decay scheme for ^{24}Na*; and b, an idealised γ spectrum expected for* ^{24}Na*. Photopeaks appear at* 1.37 *and* 2.76 MeV*, a small sum peak at* 4.14 MeV *and two escape peaks at* 2.25 *and* 1.74 MeV

6.6 Scattering effects

The features of γ spectrometry considered so far have been those which give rise to relatively sharp peaks in a γ spectrum; peaks which could be mistaken for γ photopeaks. Unfortunately there are three other signals which may appear on γ spectra and which, although readily distinguished from photopeaks, do affect the overall appearance of the spectra and can mask genuine photopeaks.

The first of these arises because of Compton scattering of γ photons passing through the detector, and is a feature most commonly associated with crystal scintillation detectors. The photopeaks observed by scintillation detectors result from photoelectric interactions of γ photons with atoms within the crystal. When Compton scattering occurs only part of the γ photon energy is converted into electron kinetic energy, and the subsequent scintillation is less intense than that which would following photoelectric absorption. The energies of the Compton electrons range from zero up to a maximum value known as the Compton edge. The effect of Compton scattering on a γ spectrum showing a single photopeak is shown in *Figure 6.15*, where other effects have been omitted.

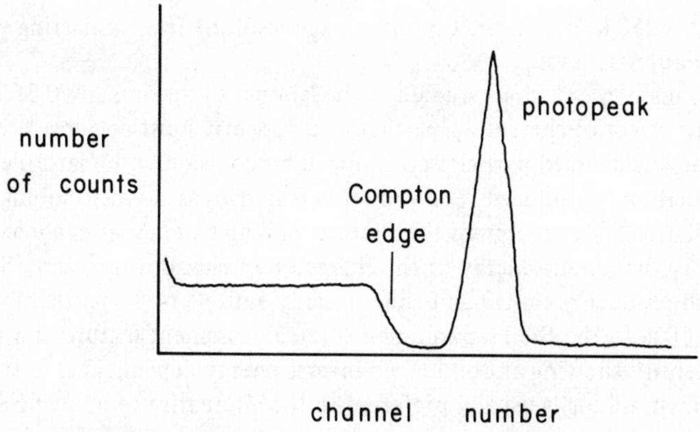

Figure 6.15 Effect of Compton scattering on a typical γ spectrum (other effects omitted)

The second effect is also a result of Compton scattering, but this time the scattering is by materials outside the detector crystal (e.g. air, the radioactive sample, support or shielding materials). Just as the electrons produced by Compton scattering have a maximum energy for a given incident γ-photon energy, so the scattered photon has a minimum energy. In fact the minimum energy of the scattered photons is when the scattering is through 180°, i.e. the scattered photons travel back along the direction from which the incident photon came. With the conventional geometric arrangement of sample and detector (shown in *Figure 6.3*), most of the γ photons which are scattered outside the detector and then enter it have been scattered through close to 180°. The result of this backscattering is a rather broad peak in the γ spectrum at energies well below the relevant γ photopeak.

This effect is shown in *Figure 6.16*, where other effects have again been omitted. With increasingly large γ-photon energies, E_γ, the backscatter peak approaches

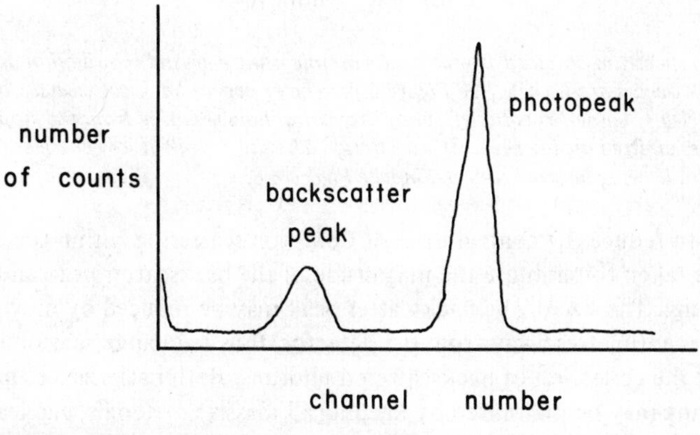

Figure 6.16 Effects of γ photon backscattering on a typical γ spectrum (other effects omitted)

an energy of ~250 keV, and the Compton edge resulting from scattering within the detector approaches $E_\gamma - 250$ keV.

Finally as many γ emissions studied in the laboratory accompany β decay processes, the effect of energetic β particles on γ spectra must be considered. When fast moving charged particles are stopped by collision with matter, electromagnetic radiation is produced. This radiation is known as Bremsstrahlung (braking radiation). The energy of the Bremsstrahlung extends from a maximum value, equal to the kinetic energy of the charged particles, down to zero. Since β decays which produce γ-emitting nuclides usually emit β^+ or β^- particles with energies > 100 keV, the Bremsstrahlung is often a prominent feature of a γ spectrum, its intensity showing a pronounced inverse energy dependence. In spite of this the Bremsstrahlung is usually preferred to the alternative of allowing energetic β particles to actually enter the detector, and when the γ spectra of β emitting samples are collected it is standard practice to place a sheet of aluminium absorber between the sample and the detector. (Appropriate thickness is usually about five times the mean range of the maximum energy β particles.)

The effects of Compton scattering, backscatter and Bremsstrahlung are shown in *Figure 6.17*, where a γ spectrum obtained from ^{137}Cs decays using a 7.5 cm NaI (T1) scintillation detector is reproduced. While in practice there is little that

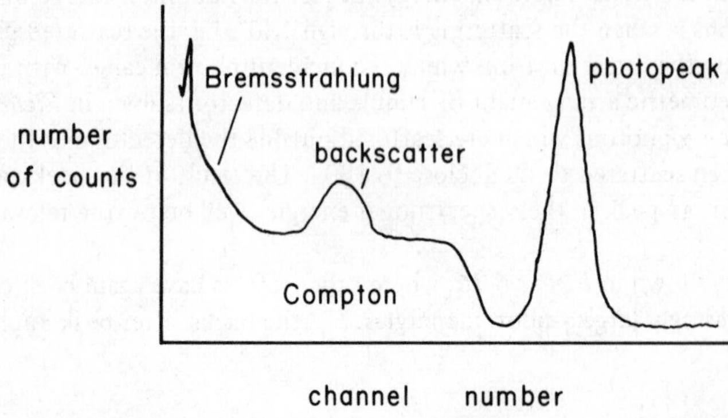

Figure 6.17 *γ spectrum obtained from a* ^{137}Cs *sample using a crystal scintillation detector. The photopeak occurs at* 662 keV *(see Figure 6.2). The effects of backscatter and Compton scattering are clearly shown as is the effect of Bremsstrahlung resulting from the stopping of the β^- particles emitted during the* ^{137}Cs β^- *decay. The sharp peak at low energies is the* Ba X *ray emitted following internal conversion (see Figure 6.8).*

can be done to reduce the contribution of Compton scattering within the detector, steps may be taken to minimise the magnitude of the backscatter peak and the Bremsstrahlung. The size of the backscatter peak may be reduced by moving the sample a few centimetres away from the detector, thus reducing the geometrical efficiency of the collection of backscattered photons. Both backscatter and Bremsstrahlung may be minimised by keeping all massive materials, e.g. lead shielding, as far away from the β emitting sample as possible.

Bibliography

Nicholson, P. W., *Nuclear Electronics*, J. Wiley & Sons, New York (1974)

Crouthamel, C. E., *Applied Gamma-ray Spectrometry*, 2nd edn, revised by Adams, F. and Dams, R., Pergamon, Oxford (1970)

Slater, D. N., *Gamma-rays of Radionuclides in Order of Increasing Energy*, Butterworths, London (1972)

7 Radiochromatography

7.1 Types of radiochromatography

Chromatographic separations of one form or another are used in virtually all chemical laboratories these days. Chromatography involving radiolabelled materials has become increasingly important in recent years, both for the separation and purification of labelled compounds and for the analysis of mixtures from tracer experiments. In this chapter some of the techniques used in radiochromatography will be described, along with a variety of detectors suitable for the different chromatographic processes.

For our purposes chromatographic separations may be regarded as falling into two classes: static and dynamic. In the first, the sample components are separated as a mobile phase moves relative to a stationary phase. The movement of the mobile phase is then terminated and this phase removed, usually by evaporation, leaving the sample components fixed in position on the stationary phase where they may be detected or collected. Classical and high performance thin-layer chromatography (TLC and HPTLC) and paper chromatography and electrophoresis fall into this class. These techniques give rise to a static distribution of components whose radioactivity may be subsequently determined. The second class is that in which the sample components are eluted from a column within a mobile phase, and detected or collected close to the point of elution before re-mixing of the components can occur. Gas, liquid and high pressure liquid chromatographies (GC, LC and HPLC respectively) are the major techniques in this class. From the radiochemical viewpoint the difference between the two classes arises through the different counting techniques required for the detection of radioactivity in a static or a flowing system.

In all kinds of radiochromatography there are two major elements which must be considered in addition to the usual factors associated with the particular chromatographic technique. These are the macroscopic size of the sample and the nature of the radionuclide involved.

The macroscopic size of a sample of radiolabelled material can be very small (*see* chapter 1, *Table 1.1*), especially when carrier free labelled compounds are involved, and yet the sample may contain a considerable amount of activity. The size of a sample may easily be so small that adsorption of components on sample applicators or injection systems represents a significant loss of sample before the chromatographic separation occurs. Small samples also have an annoying tendency to become irreversibly adsorbed on chromatographic stationary phases and support materials, thus introducing significant errors in quantitative results. Fortunately these problems are largely confined to carrier free radiolabelled materials of very high specific activities, so that widely used isotopes such as ^{14}C and, to a lesser extent, ^{3}H may normally be used in radiochromatography without much difficulty. Compounds labelled with moderately short-lived radionuclides, such as ^{32}P and ^{125}I, do require considerable care in use if the problems mentioned above are to be avoided. Addition of small (even submicrogramme) quantities of carrier, where this is possible, frequently overcomes or greatly eases problems of adsorption and the consequent sample loss. Control of the chemical state of reactive species can also be most valuable. For example, if a mixture containing $^{125}I^-$ is to be chromatographically separated by HPLC, then the addition of small quantities of a reducing agent to the eluting solvent may help to prevent oxidation of the ion within the column, which could otherwise lead to the efficient radioiodination of the column material.

At the other extreme it is often necessary to attempt radiochromatography with samples of very low specific activity, or which contain relatively large amounts of inactive material of no interest as far as the radiochemical analysis is concerned. In either case the quantity of inactive material present may seriously interfere with the desired chromatographic separation. This can result in radiochromatographic analysis being performed on the semi-preparative scale.

The second factor of major importance in radiochromatography is the type of radioactive decay of the labelled material under investigation. For example, energetic γ radiation may be relatively easy to detect with a simple scintillation detector, but it presents difficulties in thin layer chromatography because one would like to be able to resolve 'spots' of activity on the plate which may lie only a few millimetres apart. On the other hand, low energy β^- emitting isotopes are difficult to detect efficiently in most forms of radiochromatography. Several detection techniques are described below although, in view of the large number of possible combinations, the list is by no means exhaustive and only the most widely used systems are discussed.

7.2 Thin layer chromatography

Since the distribution of activity on a developed TLC plate remains static, unless the radioactive components are volatile, the procedures which have evolved for activity measurement allow counting for relatively long time-periods, so that relatively low-activity samples can be analysed without difficulty. Undoubtedly the simplest (and cheapest) procedure is to scrape narrow bands of the thin layer material into individual sample tubes, each tube then being counted for an appropriate period of time in a γ-scintillation counter (for γ emitters) or, after the addition of a suitable solvent and liquid scintillator cocktail, in a liquid scintillation counter (for virtually any other type of radionuclide). In this way a histogram can be drawn showing the distribution of activity along the TLC plate. Plate scrapers are available commercially (e.g. the Uniscil Autozonal Plate Scraper), and some scrape a band a few millimetres wide, deposit the scrapings into a sample bottle and automatically position the plate ready for the next scrape. While this is a useful procedure for the occasional radiochromatogram, it can become tedious rather quickly and may be a health hazard if used frequently with high activity samples as some of the thin-layer powder inevitably becomes airborne.

An alternative procedure involves measuring the radioactivity on the TLC plate without removing the labelled compounds or the thin layer material from the supporting plate. This may be achieved by 'scanning' the TLC plate with a detector which registers radiation passing through a narrow slit, as the slit moves slowly over the surface of the plate. One such scanning system is shown in *Figure 7.1*. For β^- emitting radionuclides a windowless G–M detector is used, so that the counter must be supplied with a continuous flow of counting gas (such

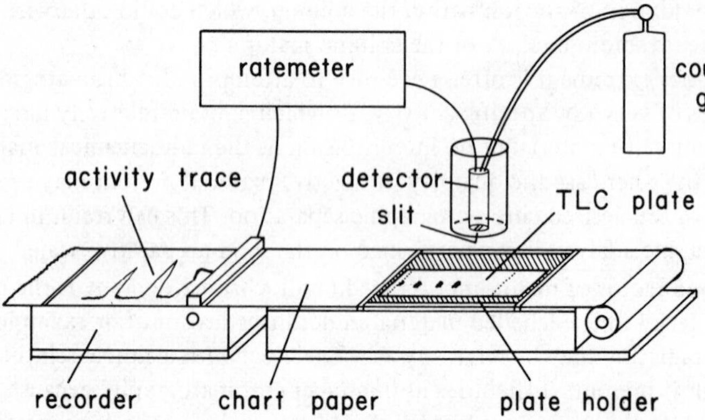

Figure 7.1 TLC radiochromatogram scanner system. For clarity the structure supporting the detector has been omitted. System shown has a gas flow detector for β^--emitting nuclides. For γ-emitting nuclides this detector may be replaced by a crystal scintillation detector

as argon/propane mixture) which escapes to the atmosphere. Because there is no window over the end of the detector even low energy β^- emitting nuclides such as ^{14}C and ^{35}S may be detected with acceptable efficiency, although the actual detection efficiency will vary with the thickness of the thin layer material as a result of β^- absorption within the layer. For γ-emitting nuclides the G–M detector may be replaced by a γ-scintillation detector which is lead shielded, except for the narrow slit which admits the γ photons. In practice this arrangement is most suitable for low energy γ emitters such as ^{125}I.

The radiochromatogram scanner shown in *Figure 7.1* performs the scanning operation by having the chart paper of the recorder move the TLC plate under the detector slit, while the count-rate is continuously recorded. This very simple system produces a chart record which is physically the same length as the TLC plate, so that the two may be placed side-by-side and spots of activity precisely located as shown in *Figure 7.2.* The precision of results recorded by this scanning technique can be varied by adjusting the speed of the chart paper movement, which in turn affects the time taken for a point on the plate to pass under the detector slit. In practice 20 cm TLC plates are scanned rapidly (1 or 2 minutes), to enable an appropriate full scale sensitivity to be selected for the count-rate, and then a chromatogram is obtained by scanning over a period of 10–60 minutes, depending on the activity on the plate.

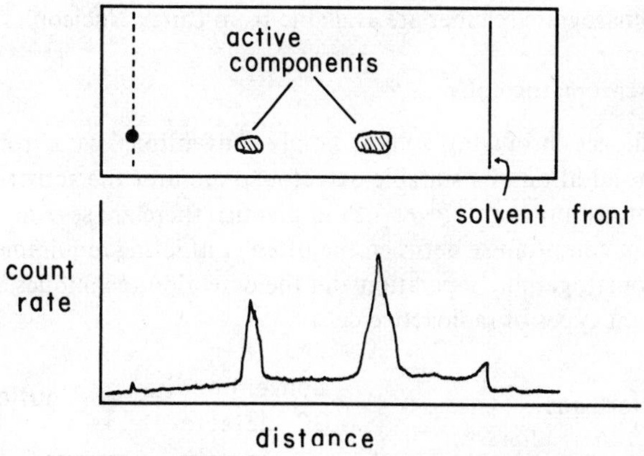

Figure 7.2 Typical radiochromatogram obtained using a scanner system, shown adjacent to the TLC plate. Small peaks at the sample application point and at the solvent front are common, and in fact can be useful for alignment of the TLC plate with the radiochromatogram

For adjustment purposes, plastic plates on which spots of ^{14}C activity have been prepared are available. Detection efficiency for any particular isotope may be determined by the use of standards of known activity spotted onto TLC plates of the type to be used in later chromatographic analysis. The detection efficiency should remain reasonably constant for any given batch of thin layer material. Typical efficiencies are up to 18% for ^{14}C and up to 3% for ^{3}H labelled materials.

Similar equipment is available for the scanning of paper chromatograms, although developments in other forms of liquid chromatography have to a large extent rendered paper chromatography obsolete, except for a few specific applications.

A relatively recent development in the field of thin layer radiochromatography is the Beta-graph, a device for photographically imaging the distribution of β^- activity on a TLC plate. The plate is positioned below two series of closely-spaced wire grids which have a high electric field applied between them in an atmosphere of argon/methane counting gas. The assembly is enclosed in a light-proof box with a Polaroid camera focussed on the wire grids. When a β^- decay on the plate gives rise to a β^- particle which passes between the grid wires, a localised spark occurs and the light flash is recorded as a small spot on the Polaroid film. After a period of time the distribution of activity on the plate is reproduced as a distribution of spots on the film. Following development an episcope within the Beta-graph is used to project a same-size image of the distribution of spots onto the TLC plate, allowing regions of activity to be precisely located.

The Beta-graph spark imaging system is intended for the location of activity on a plate so that active components may be removed for counting. It is a particularly valuable instrument for use with two dimensional radiochromatograms, which are difficult to handle otherwise, and electrophoretograms. The Beta-graph and the TLC radiochromatogram scanner are available from Panax Nucleonics.

7.3 Radiogas chromatography

In principle radiogas chromatography is simply conventional gas chromatography modified by the addition of a suitable detector to monitor the activity of the column effluent, as shown in *Figure 7.3*. In practice there are several variants which arise from compromise between the often conflicting requirements of the high speed chromatographic separation and the detection techniques associated with the different types of radioactive decay.

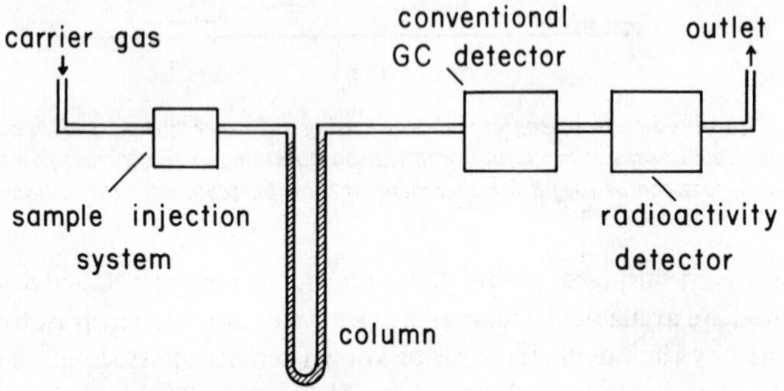

Figure 7.3 Schematic arrangement of major elements of a radiogas chromatograph

The main problem arises because most radioactivity detectors are not suitable for operation at the high temperature used for many GC separations. For example, most gas ionisation detectors cannot be operated satisfactorily at temperatures much above 100°C, and photomultiplier tubes used in γ-scintillation detectors become electronically noisy above room temperature. On the other hand, if the chromatograph effluent cools to significantly below the boiling point of an eluted component, peak tailing or even condensation within the detector may cause serious difficulties.

The detection of γ-emitting radionuclides in the effluent of a chromatograph operating close to room temperature is quite straightforward. A coil of tubing (glass, plastic or even thin-walled metal tubing) may be wrapped around the outside of a NaI(Tl) scintillation detector. Where possible a fairly large internal volume (5–20 ml) should be used to ensure that the active components are counted for a period of at least 15 seconds—although, of course, the actual volume required will depend on the gas flow rate. This normally necessitates the use of a fairly large crystal (say 7.5 cm). To minimise the high background count rate normally obtained with large crystals, a single channel analyser may be incorporated in the detecting electronics so that only the major photopeak of interest is counted. This simple arrangement must have a detection efficiency of $< 50\%$ (and often $< 10\%$ for γ-photon energies above 0.5 MeV), and if efficiency is all important, then a pair of scintillation detectors viewing a flat coil of tubing with 2π geometry is preferred. A typical 'home-made' system is illustrated in *Figure 7.4.*

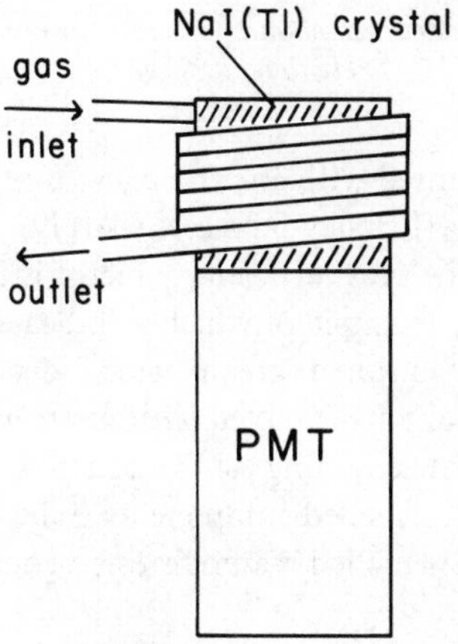

Figure 7.4 Simple γ detector for radiogas chromatography. Note that this arrangement is not particularly efficient and is unsuitable for use above room temperature. For clarity, shielding has been omitted

For operation at higher temperatures it is necessary to maintain the temperature of the flow cell, possibly by electrical heating, while keeping the scintillation detectors cool. The cooling can be achieved by clamping with a water-cooled aluminium heat shield around the detector crystal, although this will undoubtedly lower the counting efficiency.

Most β^- emitting radionuclides with decay energies greater than that of ^{14}C may be detected in flowing gases using a 2π window proportional counter, which will operate up to $\sim 100°C$. The carrier gas and eluted components flow through a rectangular cavity with thin aluminium or beryllium windows which allow β^- particles to pass into either of a pair of proportional counters. One such arrangement is shown in *Figure 7.5*, where the proportional counters are supplied with a counter gas (such as argon/methane) from a cylinder. While the presence of the window material means that ^3H-labelled compounds cannot be detected satis-

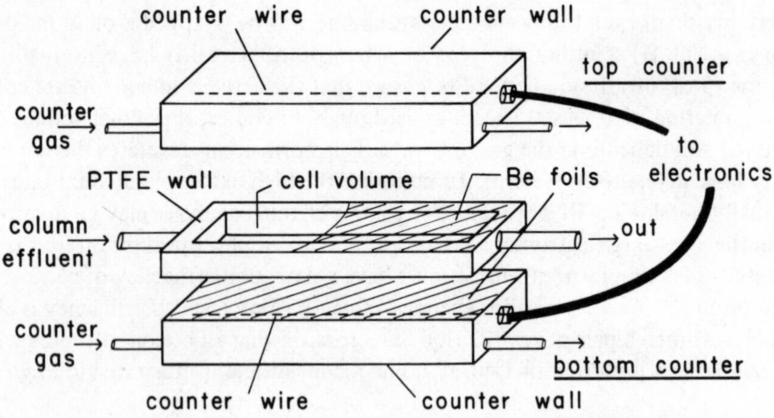

Figure 7.5 2π-'Window' proportional counter suitable for radiogas chromatography of β^- emitters with β^- energies $\geqslant$ ^{14}C. These detectors can be operated up to ~ 100°C

factorily, ^{14}C may be counted with an efficiency of ~15%, and for higher energy β^- emitters the counting efficiency increases sharply.

For tritium counting the most efficient detector is undoubtedly the gas-flow proportional counter, one example of which is indicated in *Figure 7.6*. In this case the effluent from the chromatograph–which ideally would be operating with a carrier gas of helium or argon–is mixed with methane or propane (so that this gas mixture forms a suitable counting gas) and then flows into the proportional counter, allowing any radiolabelled components to be detected. While this arrangement will allow ^{3}H-labelled materials to be counted with an efficiency

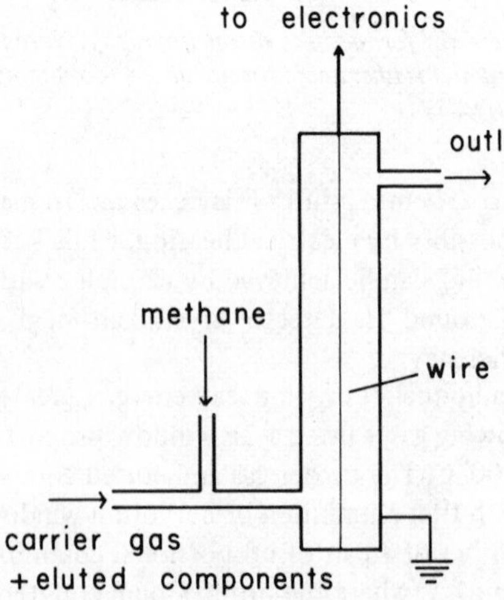

Figure 7.6 Gas-flow proportional counter suitable for radiogas chromatography of ^{3}H-labelled materials

up to ~80%, the passage of macroscopic amounts of material through the counter may alter the gas amplification factor (and hence the counting efficiency) while an active component is being counted. Unless some kind of continuous efficiency monitoring is incorporated, gas-flow proportional counters should be used only for separations in which microscopic quantities of material elute with radioactivity peaks.

An alternative method for the detection of β^- emission relies on the properties of scintillator plastics. When scintillation materials like PPO are dissolved in solvents, such as vinyl toluene, which are then polymerised, plastic scintillator materials are formed. Plastic scintillators are available commercially and many types may be machined or cast to form β^- particle detectors based on the scintillation emitted when β^- radiation enters the plastic material. For radiogas chromatography a typical detector consists of a plastic scintillator flow cell viewed by a photomultiplier tube and enclosed in a light-proof container. An example is shown in *Figure 7.7*.

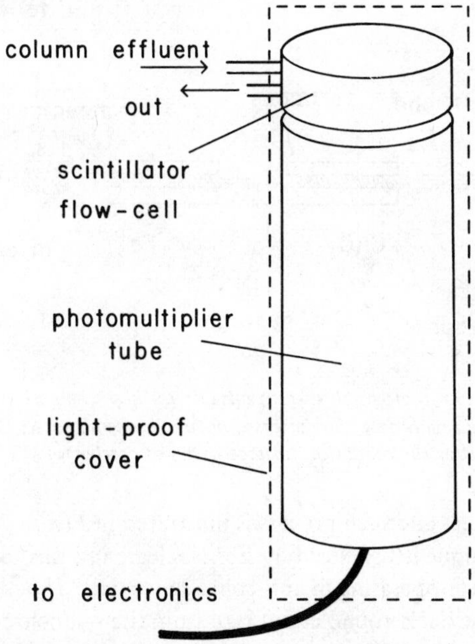

Figure 7.7 Gas-flow scintillator cell detector for radiogas chromatography of β^- emitting materials. Plastic cells are generally suitable only at room temperatures, although glass cells can be used at up to 150°C

In general plastic scintillator flow cell detectors are best suited to the detection of rather energetic β^- emitters, as the β^- particle range in a gas at atmospheric pressure becomes rather small for β^- energies below 100 keV. However, such cells are used routinely for the detection of ^{14}C-labelled compounds in systems where high detection efficiency is not essential. Several flow cells are available (e.g. from Nuclear Enterprises Ltd), as are similar cells packed with small particles of plastic scintillator or anthracene (a highly efficient organic scintillator) which are designed to compensate for the small range of β^- particles from ^3H and ^{14}C. Flow cells based on glasses containing rare-earth elements function in a similar way and are particularly useful for detection at elevated temperatures (up to ~150°C) or for compounds which may attack plastic materials. For most flow cell scintillation detectors efficiencies of 1-2% may normally be expected for ^3H, ~20% for ^{14}C and more than 50% for higher energy β^- emitters.

For high temperature radiogas chromatography of ^3H- or ^{14}C-labelled organic compounds, a particularly valuable technique involves the high temperature oxidation of labelled components, in a chromatograph effluent of argon/CO_2 carrier gas, producing ^3H_2O and $^{14}CO_2$. To prevent subsequent condensation the effluent is then mixed with hydrogen and passed over heated iron, to reduce ^3H_2O and H_2O to $^3H_2$ and H_2. The effluent is then passed into a gas-flow

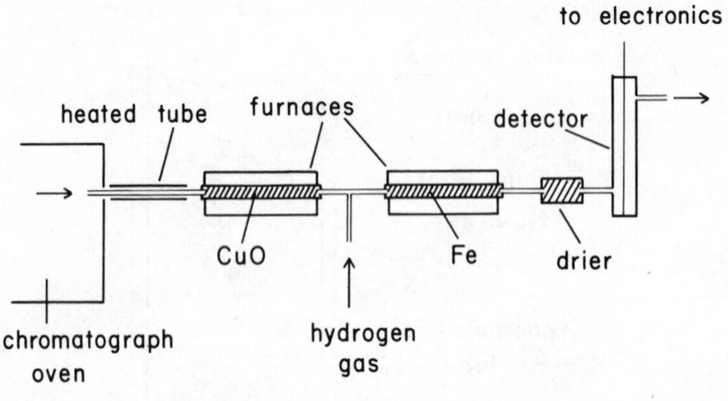

Figure 7.8 Radiogas chromatograph system suitable for high-temperature chromatography of ^3H *and* ^{14}C *labelled materials. Carrier gas is normally premixed* Ar/CO_2*; hydrogen is mixed with gas flow after the oxidation furnace to act as carrier for* HT

proportional counter: one such process is illustrated in *Figure 7.8*. One instrument based on this technique is produced by ESI Nuclear, and this uses a gas-flow proportional detector operating in anti-coincidence with a background radiation detector, so that the background count rate from the radiochromatograph detector is kept to a very low level. This instrument has a detection efficiency of more than 95% for ^{14}C and ~65% for ^3H.

7.4 Radioliquid chromatography

For liquid chromatography of radiolabelled compounds there is apparently little in the way of special equipment available commercially, although the dramatic improvements in liquid chromatography during the past few years must result in this situation changing before long. Probably the simplest way of detecting radio-

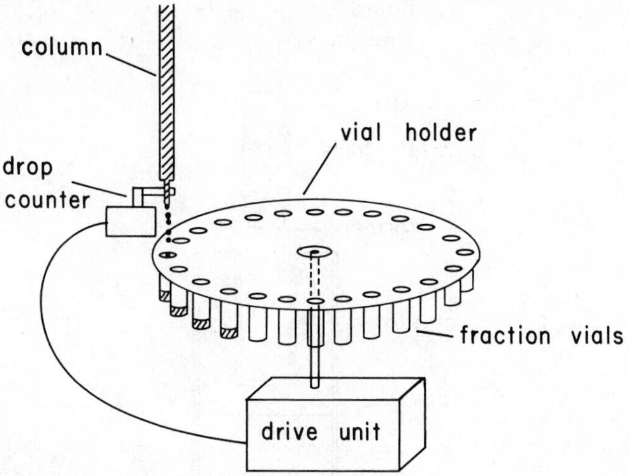

Figure 7.9 Schematic diagram of fraction collection system often used in radioliquid chromatography. Some instruments allow collection of a predetermined number of drops in each vial, others collect the drops for a preset time

active materials in the eluant of a liquid chromatograph is to collect fractions in separate bottles for subsequent counting: this technique is illustrated in *Figure 7.9*. Several fraction collectors available commercially will accept the disposable plastic vials suitable for counting in a well-crystal scintillation detector (for γ emitters) or, after addition of a few millilitres of liquid scintillator cocktail, for counting in a liquid scintillation counter (for most other classes of radionuclide).

Collecting fractions and counting when time permits, allows the experimenter to carry out radioliquid chromatography with very small amounts of activity, and using a very wide range of radionuclides. However, for certain applications—particularly those involving very short lived isotopes—flow detection may be desirable. For the detection of β^- emitting nuclides in flowing liquids there are some manual liquid scintillation counters which may be fitted with quartz flow cells filled with glass or plastic scintillator beads or anthracene crystals. One such arrangement is that found in the Coruflow system, where the flow cell is held in a sliding drawer in place of a standard liquid scintillation sample bottle.

Alternatively a purpose built detector may be constructed using a scintillator flow cell, of the kind described for radiogas chromatography, together with a photomultiplier tube. Plastic flow cells containing anthracene crystals or plastic scintillator beads (suitable for detecting β^- emitters in aqueous solvents) and glass flow cells containing CaF_2(Eu) crystals (suitable for use with organic solvents) are available commercially (e.g. from Nuclear Enterprises Ltd.). Again for ^{3}H and ^{14}C the counting efficiencies are rather low—typically 1–2% for ^{3}H and ~20% for ^{14}C—although where fairly high activity levels are involved (e.g. $<$ 10 kBq for the smallest ^{3}H-labelled component) scintillator flow cells do provide a very convenient solution to the detection problem.

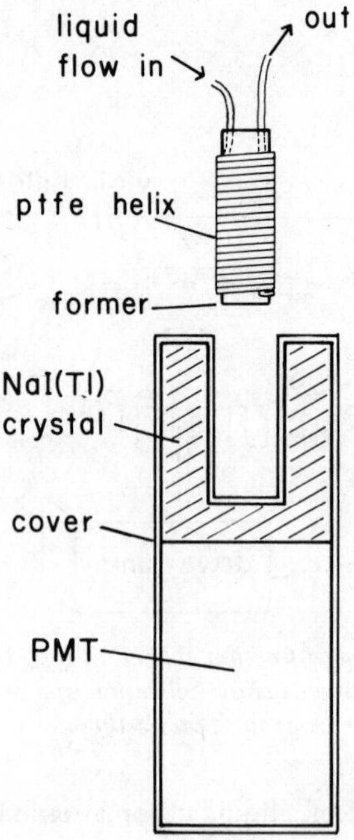

Figure 7.10 Simple flow-cell detector system for γ-emitting components in liquid chromatography. The flow cell fits in the well of a well crystal γ-scintillation detector

The detection of γ emitters in liquid chromatography effluent is particularly straightforward since a flow cell can be constructed simply by wrapping narrow bore ptfe tubing around a 5 ml plastic tube, and inserting the resulting helix into the well of a well-crystal scintillation detector. In this case the volume of the flow cell may be altered by changing the length or bore of the tubing, so that the volume may be optimised for each chromatographic separation. A typical detector is shown in *Figure 7.10*, and is expected to have a good detection efficiency for γ-photon energies between 20 keV and 1 MeV, although the actual efficiency will depend on the size of the crystal and the range of pulse heights counted.

7.5 Activity determination

For both radiogas and radioliquid chromatography the quantitative information obtained is the count rate or the integrated count as a function of time or volume of carrier eluted. In the case of fraction collection followed by fraction counting, an activity may be determined for each fraction, assuming that the counting efficiency of the detection system has been determined and that an appropriate correction for background is performed. The results may then be presented in the form of a histogram showing the variation in fraction activity with elution volume or fraction number, as shown, for example, in *Figure 7.11*. The activity of an

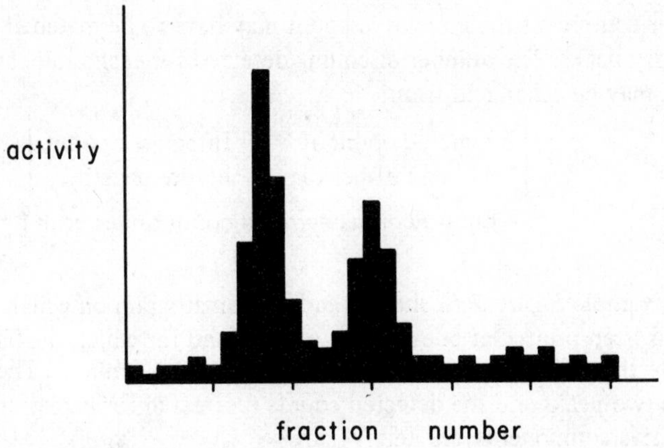

Figure 7.11 Form of radiochromatogram obtained when fraction collection technique is used

eluted component may be calculated by summing the activities for each fraction containing that component.

For flow cell detection systems the data are normally obtained in two forms as shown in *Figure 7.12*: first, a chart record showing the variation of the detected count rate with time, the radiochromatogram; second, a regular print out of the integrated count recorded since the start of the run. On modern computerised equipment this latter information can be printed directly on the radiochromatogram (as in *Figure 7.12*) and indeed subsequent calculation may be fully automat-

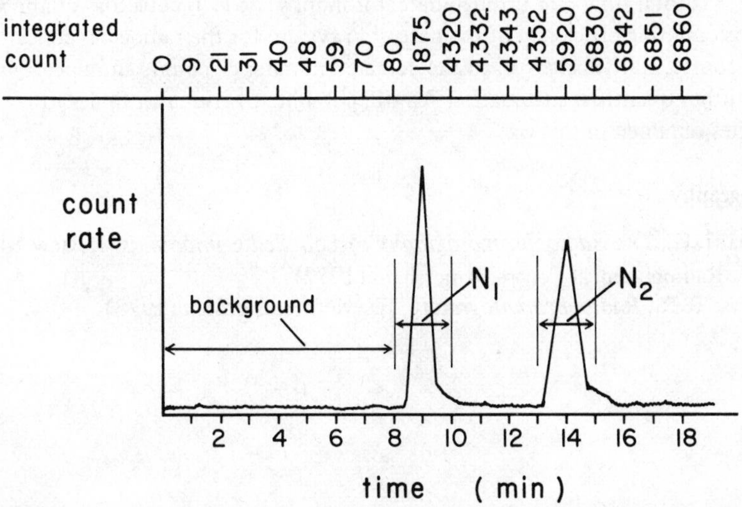

Figure 7.12 Typical radiochromatogram obtained using a flow-cell detection system. Quantitative results may be obtained by recording the integrated count at regular intervals as shown above

ed. On older equipment the integrated count may have to be matched manually to the activity peaks. The number of counts detected for each radioactive component, N_i, may be calculated from

$$N_i = \left[\begin{matrix}\text{Integrated count at} \\ \text{end of peak i}\end{matrix} - \begin{matrix}\text{Integrated count} \\ \text{before peak i}\end{matrix}\right] - \text{Equivalent background count under peak i} \tag{7.1}$$

By way of example *Figure 7.12* shows a radiochromatogram on which the integrated count has been printed at one-minute intervals and for which the background count rate (estimated from the first few intervals) is 10 ± 1 min^{-1}. The chromatogram shows two peaks and the detected counts (corrected for background) for the two radioactive components, N_1 and N_2, are

$$N_1 = [4320 - 80] - 2 \times 10$$
$$= 4220 \pm 65$$
$$N_2 = [6830 - 4352] - 2 \times 10$$
$$= 2458 \pm 50$$

Where absolute activities of eluted components are required these may be calculated from the detected count using

$$A_i = \frac{N_i \times F}{V_c \times \epsilon} \text{ Bq} \tag{7.2}$$

where F is total flow rate through detector in ml s^{-1}; V_c = detector volume in ml; and ϵ is counting efficiency of the detection system for the radiolabel concerned.

Of course, statistical errors in the detected number of counts and uncertainties in the other quantities in equation 7.2 will combine to give error limits to activities obtained in this way.

Bibliography

Sheppard, G., *The Radiochromatography of Labelled Compounds*, Review 14, The Radiochemical Centre, Amersham (1972)

Roberts, T. R., *Radiochromatography*, Elsevier, Amsterdam (1978)

8 Some Radioanalytical Techniques

8.1 Carriers and scavengers

Radioisotopes have many applications in analytical chemistry, and the ease with which even small amounts of activity may be detected has resulted in the widespread use of radioanalytical techniques for the quantitative analysis of non-radioactive materials down to the sub-picogramme level. As we have seen there are difficulties associated with the handling of carrier-free labelled compounds, largely resulting from the adsorption of such materials on apparatus, filter papers, or even dust particles within a solution. For this reason most radioanalytical work is performed with labelled materials to which chemically identical carrier materials have been added. Of course, the addition of carrier reduces the specific activity of a labelled compound, and consequently the ultimate sensitivity of the analytical technique utilising that compound, so that the amount of carrier is normally kept small in macroscopic terms, while still being orders of magnitude greater than the amount of radioactive material.

When carrier is to be added to a given activity, A Bq, of a labelled compound of specific activity S_1 Bq mol^{-1} for the purpose of producing a specific activity S_2, the weight of carrier to be added, W, may be calculated using:

$$W = AM\left(\frac{1}{S_2} - \frac{1}{S_1}\right) \text{ g} \qquad (8.1)$$

where M is the molar mass of the compound.

Conversely if W g of carrier is added to an activity, A Bq, of a compound of specific activity S_1, the resultant specific activity is

$$S_2 = \frac{S_1 A M}{W S_1 + A M} \text{ Bq mol}^{-1}$$

For example, if 0.015 g of sodium iodide ($M = 150$ g) is added to a solution containing 10 GBq of Na ^{125}I solution of specific activity 8.06×10^{16} Bq mol^{-1} (i.e. carrier free $^{125}I^-$), the resultant specific activity is

$$S_2 = \frac{8.06 \times 10^{16} \times 10^{10} \times 150}{(8.06 \times 10^{16} \times 0.015) + (10^{10} \times 150)} \approx 10^{14} \text{ Bq mol}^{-1}$$

Of course, the total activity of the sample remains 10 GBq.

The main roles of carrier are to protect against loss of radiolabelled material by physical processes, such as adsorption or volatilisation, and to reduce the risk of loss by chemical reactions which may occur with impurities which may be present in, for example, a solvent used in an analytical procedure. However, there are times when a radiolabelled reagent will need more protection than can be provided by the addition of reasonable amounts of carrier alone. For example, materials which are particularly susceptible to oxidation or reduction may have to be maintained in a particular oxidation state by the addition of small amounts of a holding oxidant or reductant. Thus aqueous radioiodide solutions, even those containing small amounts of carrier (e.g. μg ml^{-1} concentrations), are particularly prone to aerial oxidation, and may be protected by the addition of small quantities of sodium metabisulphite, provided that this will not interfere with the application intended for the radioiodide.

In some cases it becomes necessary to carry out precipitations in the presence of radiolabelled materials. Carrier free radioactive ions often undergo adsorption on such precipitates, so that it becomes necessary to use a hold-back carrier to ensure that the labelled ion remains in solution. On other occasions it becomes useful to be able to remove a radioactive impurity from a solution by deliberately encouraging adsorption of the labelled species on a freshly formed precipitate. For example, if an aqueous solution of a radioactive reagent contains some undesirable high specific-activity impurities, addition of a hold-back carrier for the reagent, followed by the formation of a precipitate of ferric or aluminium hydroxide in the solution, will usually ensure that the impurities are removed from the solution by adsorption on the precipitate—leaving the carried reagent in solution. This procedure is known as the scavenging of impurities and the hydroxide precipitates are called scavengers.

8.2 Simple analytical methods

A useful technique for determining the quantity of a material, X, in a sample is to react the sample with a radiolabelled reagent, Y^*, chosen because it undergoes a quantitative reaction with X to form a derivative, Z^*, which retains the radiolabel:

$$X + Y^* \rightarrow Z^* \qquad (8.2)$$

If an excess of the radioactive reagent, Y^*, of known specific activity (S_Y Bq mol^{-1}) is used, and if, after reaction, the labelled product Z^* can be isolated and its activity, A_Z, determined, then the original quantity of X, x can be calculated using

$$x = \frac{A_Z}{S_Y} \text{ mol} \qquad (8.3)$$

This technique is known as isotope derivative analysis and it relies on the fact that the specific activity of the derivative Z^* is the same as that of the reagent Y^* in equation 8.2. One of the best known examples of the application of isotope derivative analysis is the estimation of amino acid paper- or thin layer chromatograms by the formation of radiolabelled derivatives following reaction with radioiodinated pipsyl chloride:

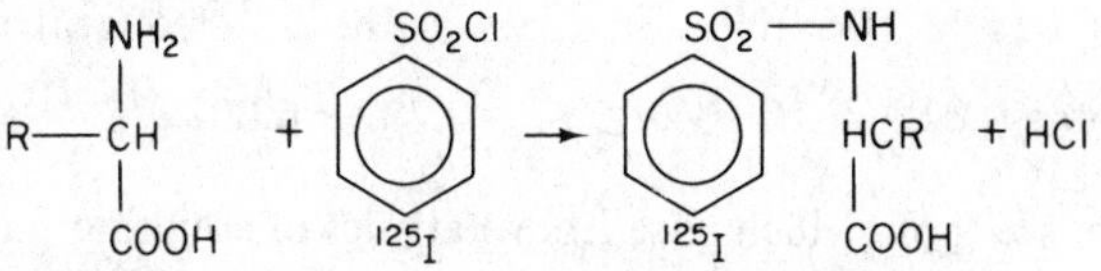

Areas of the chromatogram which contain the amino acids are made radioactive by derivative formation, and the activity measured by scanning (using a low-energy γ-photon detector) or by physically cutting up the chromatogram and counting individual sections. A worked example illustrates the calculation method typical for isotope derivative analyses.

An aqueous solution containing Zn^{2+} is provided for zinc assay. To 1 ml of this solution is added 5 ml of a solution of diammonium hydrogen phosphate (accurately standardised, 0.050 M $(NH_4)_2$ HPO_4 and pH adjusted to 6.6) labelled with ^{32}P-phosphate to a specific activity of approximately 10^7 Bq mol^{-1}. The solution is heated until a precipitate of $ZnNH_4.PO_4$ forms and begins to settle. The mixture is cooled to ~0°C and filtered through a 2 cm diameter filter paper, the precipitate being washed with inactive 0.1 M $(NH_4)_2HPO_4$ (to remove adsorbed activity). After drying, the filter paper and precipitate are placed on a 2 cm diameter counting planchet, covered with thin adhesive tape and counted using a castled G–M counter (*see Figure 3.13a*). In a 10-minute counting period the count, corrected for background, was 80 000 ± 285. 0.50 ml of the ^{32}P-phosphate reagent solution was evaporated by dryness on a 2 cm counting planchet, covered with adhesive tape and counted as above. The result in a 10-minute counting period was 480 800 ± 695, again corrected for background.

From the equation for the reaction

$$Zn^{2+} + HPO_4^{2-} + NH_4^+ \rightarrow ZnNH_4PO_4 \downarrow + H^+$$

and the two recorded counts, the concentration of Zn^{2+} in the original sample may be calculated.

The count-rate for the ^{32}P-phosphate reagent is

$$\frac{(480\,800 \pm 695)}{0.5 \times 10} = 96\,160 \pm 139 \ \text{min}^{-1}\ \text{ml}^{-1}$$

Since this solution is 0.050 M in phosphate the count rate may be expressed as

$$\frac{(96\,160 \pm 139) \times 10^3}{0.050} = (1.9232 \pm 0.0027) \times 10^3 \ \text{min}^{-1}\ \text{mol}^{-1}$$

The count-rate observed for the $ZnNH_4PO_4$ precipitate was $8000 \pm 28\ m^{-1}$, corresponding to

$$\frac{(8000 \pm 28)}{(1.9239 \pm 0.0027) \times 10^9} = (4.160 \pm 0.016) \times 10^{-6} \text{ mol (phosphate)}$$

As this amount of material derived from 1 ml of the original solution, that solution was also $(4.160 \pm 0.016) \times 10^{-3}$ M in Zn^{2+}, hence the Zn^{2+} concentration was

$$(4.160 \pm 0.016) \times 10^{-3} \times 65.37 = 0.272 \pm 0.001 \text{ g l}^{-1}$$

(Error limits shown are those which arise from statistics of counting. Error from other sources, e.g. volume measurements, have not been included.)

Isotope derivative analysis does have limitations because of the requirement that derivative formation and recovery are quantitative. As these requirements are often difficult to meet in practice, one of the isotope dilution techniques may have advantages (see below).

Closely related to isotope derivative analysis is a technique in which a chemical reaction results in the release of a radioactive material from one physical state into another. For example, the amount of water in an organic solvent may be estimated by the addition of a quantity of tritiated lithium aluminium hydride of known specific activity. Any water in the solvent reacts to form HT,

$$H_2O + \tfrac{1}{4}\,LiAlT_4 \rightarrow HT + \tfrac{1}{4}\,[LiOH + Al(OH)_3]$$

and measurement of the activity of HT released allows the quantity of H_2O to be calculated. This technique is known as radio release analysis. Another example is the estimation of certain gases by the release of ^{85}Kr from a solid form (e.g. as a quinol clathrate) following reaction with the gases in question.

Other simple analytical techniques have been modified to accommodate radioisotopic detection techniques. Among the best known are titration using a radiolabelled reagent, which involves monitoring the activity of, say, the supernatant liquid during a precipitation reaction, and polarography with radioactive metal ions, which involves collecting the mercury droplets and counting the activity they contain.

8.3 Isotope dilution analysis

The analytical techniques outlined above require that a radiolabelled product can be formed and recovered quantitatively. In practice comparatively few reactions go to completion (particularly at the very low reagent concentrations used in radioanalytical work), and most purification procedures (e.g. washing, and crystallisation) result in some loss of product. Isotope dilution analysis (IDA) is a powerful method of determining small amounts of a material, X, which is in a form where 100% separation and recovery of X are difficult. In principle a quantity of radiolabelled material in the same chemical form as X and of known specific activity, S_1, is added to the unknown quantity of inactive X. Determination of the specific activity of the mixture, S_2, then allows the unknown quantity to be calculated as described below.

Suppose we start with x_1 g of the radioactive material X^* of activity A_X and molar mass M_X. The specific activity of this material is:

$$S_1 = \frac{A_X M_X}{x_1} \text{ Bq mol}^{-1} \tag{8.4}$$

If x_2 g of inactive material X is now added the specific activity becomes:

$$S_2 = \frac{A_X M_X}{x_1 + x_2} \text{ Bq mol}^{-1} \tag{8.5}$$

In standard isotope dilution analysis x_2 is the quantity we wish to determine.

If we now perform some operation on the (labelled) material, X^*, such as a chemical reaction which yields some labelled product Y^*, i.e.:

$$X^* + U \rightarrow Y^* + V$$

then, even though we may not be able to recover all of the product Y^*, the specific activity of the recovered Y^* will still be S_2 as the ratio of radioactive atoms to equivalent inactive atoms cannot be altered by a chemical reaction. If the weight of Y^* recovered is y g, then the specific activity can be determined from:

$$S_2 = \frac{A_Y M_Y}{y} \text{ Bq mol}^{-1} \tag{8.6}$$

where A_Y is the activity of the recovered material and M_Y is the molar mass of Y.

S_2 is now known and we may rewrite equation 8.5 as

$$\frac{1}{S_2} = \frac{x_1}{A_X M_X} + \frac{x_2}{A_X M_X}$$

and using equation 8.4 this becomes

$$\frac{1}{S_2} = \frac{1}{S_1} + \frac{x_2}{A_X M_X}$$

which may be rearranged to give

$$x_2 = \left(\frac{1}{S_2} - \frac{1}{S_1}\right) A_X M_X \tag{8.7}$$

or, again using equation 8.4

$$x_2 = x_1\left(\frac{S_1}{S_2} - 1\right) \text{ g} \tag{8.8}$$

The form of equation 8.8 suggests that it is not essential for absolute activities to be determined for IDA. In practice count-rates (e.g. R_X and R_Y for x_1 g of X^* and y g of Y^* respectively) are normally used and relative specific activites (S_1' and S_2' with units of counts per unit time per mol) derived for use in equation 8.8. Not surprisingly this can only be done where R_X and R_Y are determined with identical (albeit unknown) counting efficiencies, so that care must be taken to ensure that the count rates are obtained under the same conditions (such as counting geometry and, for α or weak β^- emitters, sample thickness).

A worked example will illustrate the calculation procedure for isotope dilution analysis. (Note that errors in volume measurement have been omitted from this calculation.)

A solution containing an unknown concentration (x_2 g 1^{-1}) of disodium hydrogen phosphate was provided for analysis. To 5.0 ml of this solution was added 5.0 ml of a solution containing $x_1 = 10.0 \pm 0.1$ g 1^{-1} of Na_2HPO_4 labelled with ^{32}P to a specific activity of $S_1 = (2.00 \pm 0.05) \times 10^6$ Bq mol^{-1}, and 2 ml of 5M HNO_3. Excess ammonium molybdate solution was then added and the yellow precipitate of ammonium phosphomolybdate which formed was centrifuged down, washed with ammonium nitrate solution and dried. A quantity of this $(NH_4)_3PO_4 \cdot 12\,MoO_3$ was weighed onto a counting planchet and counted with a castled G-M counter operating with an efficiency of 26 ± 1% for ^{32}P on planchets. The weight of precipitate counted was 20.0 ± 0.2 mg, and the count recorded in a 10-minute period (corrected for background) was 2350 ± 50.

From these results the specific activity of the precipitated phosphomolybdate ($M_R = 1877$) was

$$S_2 = \frac{(2350 \pm 50) \times 1877}{10 \times 60 \times (0.0200 \pm 0.0002) \times (0.26 \pm 0.01)}$$

$$= (1.414 \pm 0.062) \times 10^6 \text{ Bq mol}^{-1}$$

From S_2 and S_1 and equation 8.7, the concentration of Na_2HPO_4 in the sample provided was

$$x_2 = (10.0 \pm 0.1) \left[\frac{(2.00 \pm 0.05) \times 10^6}{(1.414 \pm 0.062) \times 10^6} - 1 \right]$$

$$= (4.14 \pm 0.70) \text{ g } 1^{-1}$$

The form of equation 8.8 suggests that the procedure of using a known specific activity of a radiolabelled compound to determine the macroscopic amount of an inactive compound could be reversed, so that a known quantity of inactive material could be used to quantify a radioactive substance. This latter procedure is known as reverse isotope dilution analysis (RIDA) and is often used for the determination of activity of a particular radioisotope, or for quantifying a specific labelled impurity in a radiochemical. In such cases the macroscopic quantity of the labelled material in the sample can often be neglected, so that the activity in the sample (from equation 8.5) is simply

$$A_X = \frac{S_2}{M_X x_2} \text{ Bq}$$

where S_2 is the specific activity determined for the mixture of active and inactive materials, in Bq mol^{-1}; and x_2 is the amount of inactive material used, in grammes.

Reverse isotope dilution analysis is frequently applied to the analysis of a labelled compound in a mixture containing other radioactive materials, so that it is essential that unwanted activities are not allowed to interfere with the determination of S_2. For example, in simple precipitation procedures a preliminary scavenging operation may be carried out to ensure the removal of active species which could otherwise become adsorbed on the precipitate.

8.4 Radioimmunoassay

Radioimmunoassay (RIA) is an elegant technique for the measurement of very low concentrations (typically 10^{-6}-10^{-12} g ml^{-1}) of specific compounds in the presence of large excesses of other materials. It is frequently applied to the estimation of physiological, toxicological or therapeutic agents in body fluids, plasma or serum. RIA is based on the reversible reaction between antigens and antibodies. An antigen is a substance which stimulates an immune response when injected into an animal, and is usually a high molecular weight compound such as a protein or polysaccharide. Other substances (such as hormones or drugs) may often be chemically bonded to high molecular weight carriers so that RIA procedures can be used for their analysis. Antibodies are immunoglobulins formed in an animal by the immune response system when an antigen is detected.

An antibody normally demonstrates a very high specificity in its reactions for the antigenic compound which stimulated its formation. It is this specificity that enables an antibody to be used for the selective detection of a particular antigen in the presence of many other materials, which may themselves have antigenic properties. For example, the antibody produced in a rabbit by the injection of human insulin may be collected and used for the assay of insulin in human serum without showing significant reactivity towards other serum proteins and polypeptides.

The reaction between an antigen, AG, and an antibody, AB, may be written as

$$\text{AG} + \text{AB} \rightleftharpoons \text{AG}\cdot\text{AB}$$

the product being an antibody-antigen complex. In practice this complex can usually be separated from excess antigen by centrifugation, filtration or some form of chromatography.

In the radioimmunoassay of a particular antigen a quantity of the same antigen labelled with a suitable radionuclide is allowed to compete with the unlabelled antigen for the antibody. The procedure is to take a solution containing a known amount of radiolabelled antigen, AG*, and add to it an amount of antibody which will bind with about half of the radiolabelled material, i.e.:

$$2\,AG^* + AB \rightarrow AG^* + AG^*{\cdot}AB$$

After the separation of the complex from the solution the ratio of activities is

$$\frac{[AG^*{\cdot}AB]}{[AG^*]} = 1.0$$

This procedure is then repeated with a measured volume of the solution containing the unknown concentration of the (unlabelled) antigen added. The ratio of activities in the solution (AG*) and complex (AG*•AB) may now be found from the equation

$$2\,AG^* + x\,AG + AB = \frac{2}{2+x}\,AG^*{\cdot}AB + \frac{x}{2+x}\,AG{\cdot}AB + \left(2 - \frac{2}{2+x}\right)AG^* + \left(x - \frac{x}{2+x}\right)AG \quad (8.9)$$

where x is the unknown number of molecules of unlabelled antigen.

From equation 8.9

$$\frac{[AG^*{\cdot}AB]}{[AG^*]} = \frac{\dfrac{2}{2+x}}{2 - \dfrac{2}{2+x}}$$

so that

$$x = \frac{[AG^*]}{[AG^*{\cdot}AB]} - 1 \quad (8.10)$$

If the weight of AG* used, W_{AG^*}, is known, then the weight of AG in the sample, W_{AG}, may be calculated from

$$W_{AG} = \frac{x}{2}\,W_{AG^*} \quad (8.11)$$

In practice it is usually difficult to determine the absolute amount of a labelled antigen because relatively high specific activity labelled antigens are used in RIA procedures, and a certain amount of labelled impurity is normally present, which contributes to the activity but which generally does not bind with the antibody. RIA is therefore normally performed by comparing the amount of labelled antigen bound to the antibody when the unknown quantity of antigen is present with the amounts bound in the presence of a series of standards containing known concentrations of antigen. This procedure gives a series of results for the activity of AG*•AG which may be plotted against the antigen concentrations of the reference standards to produce a dose-response curve. In practice there are several ways of plotting this calibration graph, but a typical example is shown in *Figure 8.1*, where the ordinate is the percentage of the total activity found in the antigen–antibody complex (% bound) and the abscissa is the concentration of

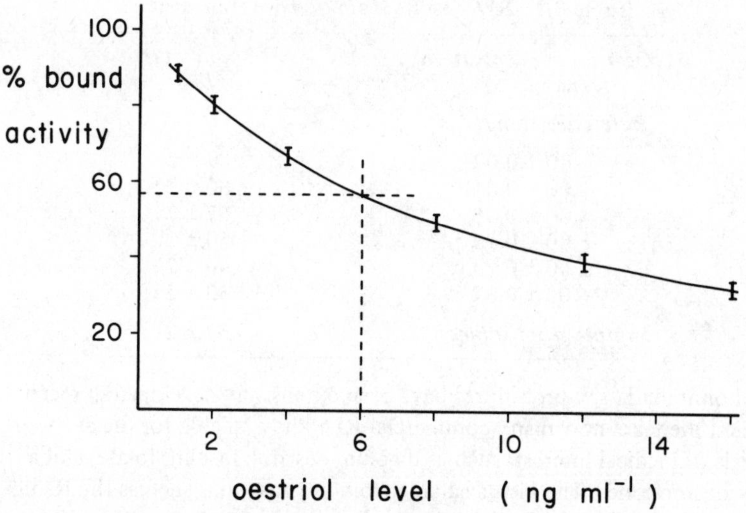

Figure 8.1 Typical radioimmunoassay dose-response curve. Points with error bars represent response from standards provided for calibration, the broken lines show how the curve can be used to determine the antigen level in a sample provided for assay. The curve shown is that discussed in the worked example of an oestriol RIA

antigen in the reference standards. The concentration of the antigen in the sample provided for assay is now simply read from the dose-response graph using the % bound activity observed for the sample.

A worked example will illustrate this calculation procedure. A sample of blood was taken from a subject during late pregnancy for the determination of unconjugate oestriol level in the serum. 20 μl samples of serum and the oestriol standards provided were pipetted into sample tubes, all tubes being duplicated. 200 μl aliquots of ^{125}I-labelled oestriol and 200 μl of anti-oestriol serum were then added to each tube and the mixtures incubated for 45 minutes at room temperature. At the end of the incubation period 500 μl of ammonium sulphate solution was added to each tube, the tubes were centrifuged for 20 minutes at 1,500 g and the supernatants discarded. The precipitates were then counted in a γ-counter, along with one tube containing 200 μl of the ^{125}I-labelled oestriol solution. The results were obtained as % bound activity.

$$= \frac{\text{activity of precipitate}}{\text{activity of 200 } \mu\text{l of solution}} \times 100\%$$

and the means for the duplicated samples are given in *Table 8.1*.

A dose-response calibration graph was drawn using the results from *Table 8.1* and is shown in *Figure 8.1*. From this curve the subject's unconjugated serum oestriol level was found to be 6.0 ± 0.4 ng ml^{-1}, where the error limits have been estimated from the graph using the standard deviation determined for the % bound activity obtained for the sample from the subject.

Table 8.1 RIA results from oestriol standards

Oestriol concentration (ng ml^{-1})	*% bound activity* (%)
Reference standards;	
1.00 ± 0.02	89 ± 2
2.00 ± 0.04	80 ± 2
4.00 ± 0.08	67 ± 2
8.00 ± 0.16	50 ± 2
12.00 ± 0.24	40 ± 2
16.00 ± 0.32	33 ± 2
Sample from subject	57 ± 2

Radioimmunoassay procedures have been extensively developed in recent years and there are now many commercial RIA kits available for the assay of materials of medical interest, such as digoxin, oestriol, insulin, folate, and a variety of hormones. The kits, available from organisations such as the Radiochemical Centre and New England Nuclear, contain a labelled antigen, the antisera, reference standards of unlabelled antigen and relatively simple instructions for performing the assay and for plotting a suitable graph for the determination of the antigen level in the sample.

Most commercial kits are based on ^{3}H-labelled or ^{125}I-labelled antigens, although other radiolabels have recently come into use in RIA procedures. In most cases a liquid scintillation counter may be used for RIA counting although several companies now offer low-energy γ-photon counters specifically designed for RIA with ^{125}I or ^{57}Co labelled materials. Many of the modern microprocessor-controlled liquid scintillation counters and γ counters have facilities for counting a series of samples and presenting the results in the form of % bound activity (or some related form), and some are programmed to plot a calibration graph and to calculate the antigen concentration in the assayed sample.

While the techniques for producing antisera are outside the scope of this book it is worth noting that RIA and related procedures can provide sensitive methods of determining a wide range of materials even where commercial kits are not available. The relative ease with which many compounds may be labelled with ^{3}H or a halogen radioisotope suggests that RIA will be an analytical procedure of increasing importance for many years to come.

Bibliography

McKay, H. A. C., *Principles of Radiochemistry*, Butterworths, London (1971)

Duncan, J. F. and Cook, G. B., *Isotopes in Chemistry*, Clarendon, Oxford (1968)

Lambie, D. A., *Techniques for the use of radioisotopes in analysis*, E. & F. N. Spon, Ltd, London (1963)

Freeman, L. M. and Blayfox, M. D. (eds), *Radioimmunoassay*, Grune & Stratton, New York (1975)

Pasternak, C. A. (ed.), *Radioimmunoassay in Clinical Biochemistry*, Heyden & Son, Ltd, London (1975)

9 The Preparation of Labelled Compounds

Volumes have been written on the preparation of radiolabelled compounds, and some—particularly the work by Evans on tritium and its compounds—provide outstanding sources of information on the subject. In this chapter only a brief outline of some of the most widely used methods for small scale laboratory labelling will be given. The intention is to provide some understanding of the kind of techniques used in radiolabelling, and to demonstrate that the procedures are not necessarily difficult to carry out. Some of the labelling methods described, involve the use of high activities of radionuclides and should not be attempted by the inexperienced without expert supervision.

9.1 Choosing the radiolabel

When a particular project calls for the use of a radiolabelled compound, the first point to settle is the choice of the radionuclide. For many organic chemical applications, the lack of any reasonably long-lived radioisotopes of oxygen or nitrogen limits consideration to the radioisotopes of hydrogen and carbon, with occasional uses for sulphur, phosphorus or one of the halogens. This restriction of choice is normally forced on the experimenter because he must choose the radiolabel from the isotopes of the elements normally present in the compound he wishes to label. However, in some applications one may be able to accept the presence of a radioisotope of an element foreign to the normal (non-radioactive) structure of, for example, a protein molecule that must be radiolabelled. Of course, if the object of radiolabelling is to trace the behaviour of one particular atom within a compound then there is no choice of the chemical identity of the radionuclide.

The second step, which may be more or less intimately related to the first, is to decide on the desirable nuclear characteristics of the radionuclide. For example, if the radiolabel is to be detected while present in a macroscopic structure, such as an animal or an item of equipment, then a γ-emitting radionuclide is probably required, although a β^+ emitter could be used, in which case the 0.511 MeV annihilation photons would be detected. On the other hand, if a radiolabelled material is to be used in experiments involving considerable laboratory manipulation, then the choice of a low-energy β^- emitter (such as ^3H or ^{14}C) may present less difficulties from the point of view of radiation shielding.

The half-life of the radiolabel may have to be considered. Short-lived radionuclides can be used to achieve very high specific activities during an experiment, and yet not present a disposal problem once the experiment is over. Indeed for some applications, such as tracing the flow of material in an industrial plant, the use of a short-lived radionuclide is virtually essential if expensive decontamination procedures are to be avoided. On the other hand, long-time scale experiments naturally require longer lived radioisotopes.

Having decided on the desirable properties of a radiolabel for a particular application, one is usually forced to compromise on at least some of the points because of the limitations of the available radionuclides. For example, there is no γ-emitting isotope of hydrogen, and the only γ-emitting isotope of carbon with a useful half-life is ^{11}C ($t_{1/2}$ ~ 20 minutes) which is hard to come by unless one has access to a cyclotron. *Table 9.1* shows some of the more readily available radionuclides together with their most important radioactivity properties.

Table 9.1 Readily available radionuclides for laboratory synthesis

Nuclide	$t_{1/2}$	*Principal emission**	*Typical chemical form*
^3H	12.5 yr	6 keV β^-	^3HHO, $^3H_2$
^{14}C	5730 yr	50 keV β^-	$^{14}CH_3OH$; $^{14}CH_3I$; $H^{14}CHO$
^{32}P	14 d	0.7 MeV β^-	Orthophosphate in dilute HCl; $^{32}PCl_3$, $^{32}PCl_5$
^{35}S	87 d	50 keV β^-	Sulphate in aqueous solution; $H_2{}^{35}SO_4$
^{36}Cl	3×10^5 yr	250 keV β^-	aqueous $H^{36}Cl$
^{82}Br	36 h	100 keV β^- 0.78 MeV γ	KBr/K ^{82}Br (low specific activity)
^{125}I	57 d	35 keV γ	$^{125}I^-$ in aqueous NaOH
^{131}I	8 d	180 keV β^- 360 keV γ	$^{131}I^-$ in aqueous NaOH

*Energies for β^- particles are mean energies

Once the radiolabel has been selected the problem reduces to one of obtaining the desired labelled compound at the desired specific activity. Although some radiolabelling techniques are described below, the most convenient method is often the direct purchase of the labelled material from one of the radiochemical suppliers such as the Radiochemical Centre. Even if a particular labelled compound is not specified in a supplier's catalogue, it is often possible to obtain the material by special order. For specialised syntheses, particularly of high specific activity materials, it may be uneconomic to attempt preparations in small laboratories which do not possess the required expertise and facilities.

9.2 ^{3}H-labelled compounds

A very wide range of compounds—mostly organic and biological materials—can be fairly readily labelled with tritium using a variety of different labelling techniques. One of the earliest ^{3}H-labelling procedures of wide applicability was the Wilzbach method. In Wilzbach labelling, the compound to be labelled, the target compound, is placed in contact with an excess of tritium gas ($^3H_2$ or T_2) and left for a time ranging from a few days to many months. An attractively simple method, Wilzbach labelling has several disadvantages which ensure that it is now only used as a last resort. The specific activity of the desired product is usually low, and yet this product frequently contains high specific activity labelled impurities which are difficult to remove.

Wilzbach labelling occurs by an ion-molecule mechanism induced by the ionising effects of the tritium β^- decay. The result is usually the isotopic replacement of an H atom by a T atom, for example,

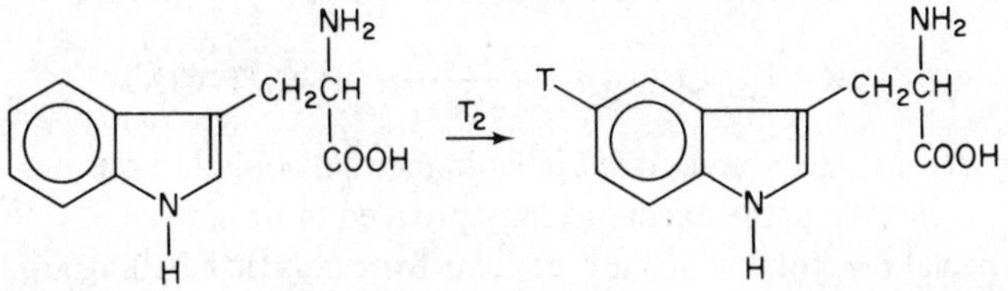

Labelling occurs more or less randomly at all available positions, so that it is not generally possible to produce specifically labelled molecules using this technique.

Various improvements have been introduced in recent years for systems in which this general method is still useful. For example, microwave or electric discharge may be used to increase the rate of ion formation, and hence the rate of labelling. Intimate mixtures of the target compound with charcoal or a noble metal catalyst have also been used to improve the specific activity of the labelled product. However, the method is not attractive for use in the research laboratory and has largely been surpassed by labelling in solution.

For the preparation of ^{3}H-labelled compounds in the laboratory there are basically two widely used techniques: (1) the catalytic exchange of one atom of the target compound for a ^{3}H atom; and (2) the chemical synthesis of the required material from another readily available ^{3}H-labelled compound. For the general preparation of tritiated compounds at moderate specific activities ($\sim 10^{11}$-10^{12} Bq mol^{-1}) catalytic exchange in solution provides a very convenient and fairly rapid labelling technique. For very high specific activity and carrier free labelled compound production, for the preparation of specifically labelled materials and for labelling of compounds which may be unstable under catalytic exchange conditions, chemical synthesis of the required compound may be more suitable.

(1) Catalysed isotopic exchange

Catalytic exchange is most commonly applied to hydrogen isotope exchange. In this case the target compound is dissolved in a tritiated solvent in the presence of a catalyst–either homogeneous or heterogeneous–and time allowed for isotopic exchange to occur. The solvent and catalyst are then removed leaving the required tritiated material. Note that in many cases some hydrogen isotopic exchange may occur without the aid of a catalyst. Exchange into labile hydrogen positions such as H atoms attached to N, O or S atoms, occurs readily, e.g.

$$CH_3COOH + HTO \rightleftharpoons CH_3COOT + H_2O$$

However, molecules labelled in this way are rarely of use as tracers as the label tends to be lost from the molecule as easily as it went in.

Exchange into non-labile positions is normally insignificant, and yet in the presence of a suitable catalyst may become quite rapid. Various catalysts have been used, including simple acids or bases, certain metal ions, aluminium trichloride, phosphoric acid-boron trifluoride reagent and a range of heterogeneous platinum and palladium metals. For example, refluxing a carboxylic acid (normally as the potassium salt) for a few hours in the presence of KOH in HTO solution results in labelling at the carbon α to the carboxyl group,

$$\text{R-CH}_2\text{-COO}^- \xrightarrow[\text{reflux}]{\text{KOH/HTO}} \text{R-CHT-COO}^-$$

Similar procedures may be used for labelling amino acids and purines, although with these compounds the exchange may proceed more slowly.

Two 'special reagents' which are popular for catalytic labelling are tritiated water containing aluminium trichloride (typically ~2M), which is thought to act through the acid $H[AlCl_3OH]$, and tritiated water containing phosphorous pentoxide through which an equivalent amount of boron trifluoride has been bubbled and which acts through the complex $TH_2PO_4BF_3$. Tritiation using aluminium trichloride is relatively straightforward and can produce compounds at moderately high specific activities within a few hours. However, in most cases the exchanging mixture requires heating if the reaction time is to be short enough to avoid radiation damage to the target compound and/or the product. $TH_2PO_4BF_3$ is used at room temperature and is therefore particularly useful for labelling heat-sensitive compounds. Unfortunately the complex is not particularly stable under the effect of β^- radiation and so does not keep well. Also it tends to attack glassware, so when being stored for short periods, polythene containers are preferred.

Both these special reagents are useful for labelling a wide range of hydrocarbons, halogen substituted hydrocarbons, alcohols, ketones and heterocyclic compounds. However, some classes of compounds are not suitable for labelling in the presence of these acidic catalysts; for example, basic compounds may react to give salts, thus preventing exchange altogether. Furthermore the $TH_2PO_4BF_3$ complex will dehydrate compounds which are susceptible to dehydration, such as carbohydrates, and so is not suitable for use in the labelling of such materials.

Catalysis of hydrogen isotope exchange using metal ions has come into increasing use since the late 1960s. In general the target material is heated for a few hours in the presence of a tritium source, such as HTO, potassium chloroplatinate, acetic acid and a trace of HCl. Again the method is restricted to compounds which are not acid sensitive, and in practice the most frequently quoted examples of materials labelled using homogeneous metal ion catalysis are aromatic compounds.

Undoubtedly the most versatile method for tritiating organic compounds in the research laboratory is the exchange reaction catalysed heterogeneously by platinum or palladium metal. The catalyst is normally used in the finely divided form obtained by reduction of the metal oxide. As in the other methods, the source of tritium is usually HTO, although tritiated acetic acid (carboxyl-T, i.e. CH_3COOT) may be used. Labelling may be achieved by leaving the target compound, the tritium source and the catalyst in contact for a few hours. In general more efficient labelling is obtained by heating the mixture, and temperatures in the range 50-100°C are normally used.

Heterogeneous catalysis is attractive for isotopic exchange labelling because of the variety of compounds which may be labelled with ease–including aromatic hydrocarbons, alicyclic compounds, amino acids, purines, pyrimidines, nucleotides and steroids–and because the labelled product is rather easily separated from the catalyst and tritiated solvent, for example, by filtration and vacuum distillation respectively. Some laboratories maintain an apparatus permanently designated for tritium labelling and which may contain an activity of several thousand GBq (1GBq = 10^9 Bq) of tritiated water. Not surprisingly it is essential that such set-ups are housed in fume cupboards especially designed (and approved) for coping with volatile radioactive material.

While catalysed isotopic exchange provides a comparatively straightforward method of labelling organic molecules it does have some limitations. For example, most exchange reactions yield products which are generally labelled, i.e. the ^3H atoms are distributed amongst all the original H atoms. The specific activity of the labelled product can be controlled to some extent by the choice of the specific activity of the tritium source, but there are limits to this. Carrier free T_2O has a specific activity of $\sim 98 \times 10^{12}$ Bq mol^{-1} and is virtually unmanageable; it glows brightly and undergoes rapid decomposition under the influence of the energy deposited by its own β^- decay. Tritiated water may be obtained from the Radiochemical Centre at a specific activity of ~ 1850 GBq mol^{-1}, and higher specific activities are available on special request.

A second catalytic exchange procedure to a large extent overcomes the non-specificity problem associated with hydrogen isotope exchange. This procedure is the catalysed exchange of a T atom for a halogen in a halogenated molecule, e.g.:

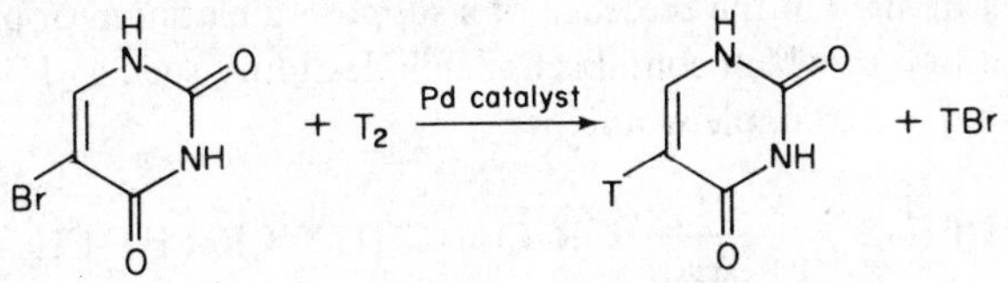

In this case the source of tritium is normally T_2 gas and the catalyst is palladium supported on charcoal. The reaction is performed in the presence of a base to remove the T-halide product.

While the specificity of catalysed T-halogen exchange is generally very high ($>$98%) it should be noted that the formation of HTO during the reactions, and the presence of the catalyst, may result in small amounts of hydrogen isotopic exchange labelling. Compounds containing any of the halogens may be used for exchange labelling, although in practice iodinated materials are usually avoided because of the tendency of iodides to poison the catalyst.

(2) Direct chemical synthesis

Where the chemical synthesis of an ^{3}H-labelled compound is unavoidable there are two basic approaches which may be adopted. The first involves the relatively minor modification of an ^{3}H-labelled compound which is already available commercially. This is undoubtedly the simpler approach and would normally be considered before the alternative. Virtually any standard chemical procedure may be adapted for the modification of commercially available tritiated compounds, although the difficulty of performing reactions on a microscale tends to limit the usefulness of this approach to the production of low to moderate specific activity compounds. For example, tritiated aldehydes may be prepared by the oxidation of tritiated primary alcohols or *cis*-diols, or from tritiated alkyl halides or tosylate by heating for a few minutes in $NaHCO_3$/DMSO:

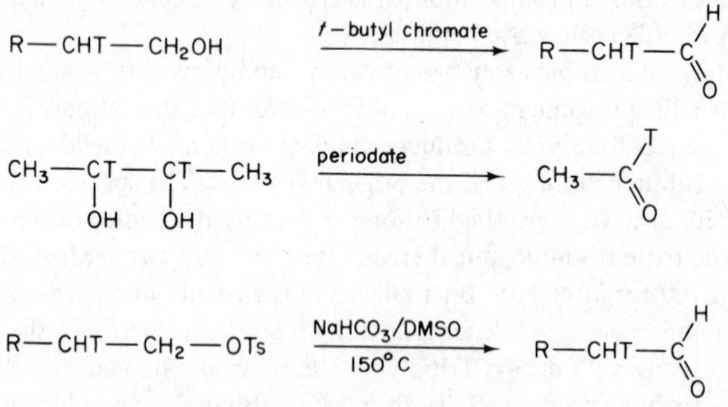

The second approach consists of labelling a compound with a tritium atom or small group using a small tritiated molecule as the tritium source. Where gas handling facilities are available, the catalytic hydrogenation of unsaturated compounds provides a relatively straightforward synthetic route. Tritium gas or T_2/H_2 mixtures are used in the presence of a supported platinum or palladium catalyst and in a solvent which contains no labile H-atoms, such as THF, dioxan, and heptane. A typical example would be

$$\text{R-CH=CH-X} \xrightarrow[\text{Pd catalyst}]{H_2/T_2} \text{R-CH}_2\text{-CHT-X} + \text{R-CHT-CH}_2\text{-X}$$

etc.

where R = alkyl group, X = halogen atom.

Reduction of carbonyl groups using tritiated metal hydrides, such as lithium aluminium tritide or sodium borotritide, also provides a convenient laboratory labelling route, e.g.:

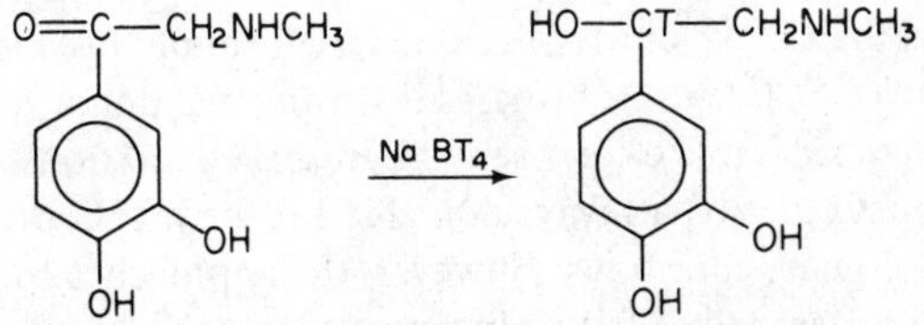

Lower specific activities may be obtained by refluxing the target compound in THF with a borohydride in the presence of HTO as the tritium source.

Grignard reagents are often useful in the synthesis of labelled compounds. Tritiated Grignard reagents may be prepared from readily available tritiated compounds and are most valuable for the incorporation of labelled small groups into larger molecules, e.g. (using the conventional Grignard formula):

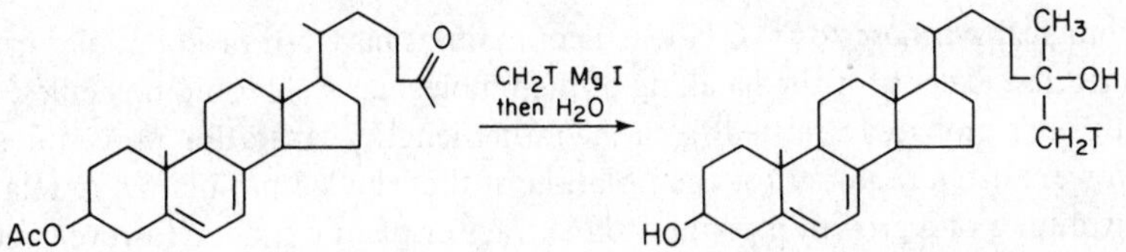

A range of biochemical labelling methods is also available for the preparation of ^{3}H-labelled compounds and many of these are described in detail by Evans (see bibliography). However, for most applications exchange labelling and chemical synthesis are straightforward enough for sophisticated biochemical procedures to be unnecessary.

9.3 ^{14}C-labelled compounds

The laboratory preparation of ^{14}C-labelled compounds is almost entirely limited to chemical or biochemical synthesis, as there is no technique comparable to the isotope exchange procedure described above for ^{3}H-labelling. The prime source of ^{14}C for labelling is $^{14}CO_2$ and the carbonation of Grignard reagents does provide a direct route for the synthesis of ^{14}C-labelled carboxyl compounds, e.g.:

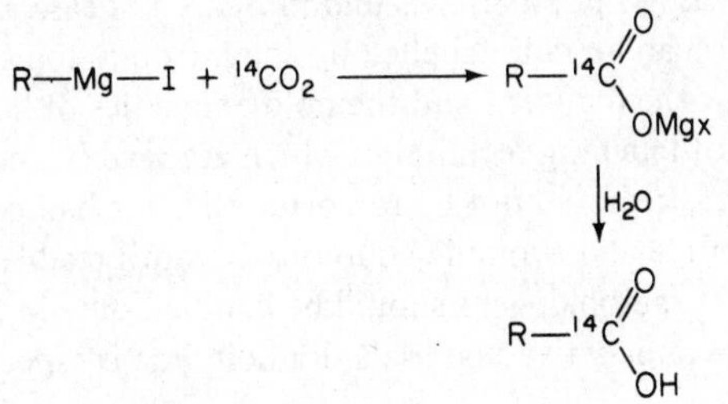

Apart from this kind of reaction, $^{14}CO_2$ is of little importance for other syntheses carried out in research laboratories. Such a wide range of ^{14}C-labelled compounds is available from radiochemical suppliers that synthesis is much more likely to involve the modification of a commercially available ^{14}C labelled material.

^{14}C-labelled compounds are more amenable to conventional small scale chemical manipulations than most shorter-lived material, simply because of the lower specific activity of ^{14}C. Even so syntheses involving maximum specific activity ^{14}C may require considerable expertise in microscale operations.

The number of synthetic pathways available for the preparation of ^{14}C-labelled compounds is, of course, enormous. However, the approaches to chemical synthesis can normally be regarded as falling into two categories, in much the same way as the synthesis of ^{3}H-labelled compounds: the modification of a commercially obtainable ^{14}C-labelled material; and the addition to an unlabelled precursor of a ^{14}C-labelled small group. Techniques used in both categories are described in detail by Murray and Williams. Preparations which fall into the second category generally utilise $^{14}CH_3I$, $^{14}CH_3OH$, $K^{14}CN$ or $H^{14}CHO$ as the source of radiocarbon, although ^{14}C-carbonates and bicarbonates are readily available for syntheses involving $^{14}CO_2$.

Biological syntheses of ^{14}C-labelled compounds have not been popular in small laboratories, except for the labelling of high molecular weight biomolecules. In part this has probably resulted from the rather lengthy separation procedures which were often required for the isolation of the labelled product from relatively large volumes of a growth medium, cell extract or plant extract. However, recent developments in ultrafiltration and high performance liquid chromatography have changed this position quite dramatically. The increasing availability of high (enzymatic) activities of enzymes and facilities for enzyme immobilisation suggests that biochemical methods for synthesising ^{14}C-labelled compounds will become more important.

9.4 Radiohalogenated compounds

The preparation of compounds labelled with radioisotopes of the halogens is the only remaining radiochemical area where it is possible to generalise. Compounds labelled with radioiodine, or to a lesser extent with radiobromine or radiochlorine, are becoming increasingly useful in a variety of research and development areas, in routine analytical applications and in nuclear medicine. (Fluorine will not be considered here as the longest lived radioisotope, ^{18}F, has $t_{1/2} \sim 2$ hours). One reason for this increase in popularity is undoubtedly the ease with which a wide range of compounds can be radiolabelled by the introduction of a radiohalogen. The monovalency of the halogens and the mild energetics of halogen chemistry give rise to a range of labelling techniques which are readily employed in a small laboratory. However, it should not be forgotten that the halogens themselves are volatile, and that iodide and bromide aquoions are both readily oxidised to a volatile form, so that radiohalogens should be handled only in fume cupboards designed for volatile radioactive materials. Radioiodine is especially hazardous because if it does enter the human body it is rapidly concentrated in the thyroid gland where it can cause serious and permanent damage.

Some of the most widely used isotopes of the halogens are shown in *Table 9.2*, where a broad selection of nuclear properties is encompassed. However, the range

Table 9.2 Common isotopes of the halogens

Isotope	$t_{1/2}$	*Principal emissions*	*Comments*
^{34m}Cl	32 min	β^+, 2.14 MeV γ	Fast neutron source required
^{35}Cl	Stable, natural abundance 75.5%		
^{36}Cl	3×10^5 yr	150 keV β^-	Commercial suppliers
^{37}Cl	Stable, natural abundance 24.5%		
^{38}Cl	37 min	4.8 MeV β^-, 2.2 MeV γ	Commercial suppliers or thermal neutron source
^{77}Br	58 h	β^+, 0.52 MeV γ	Cyclotron produced
^{79}Br	Stable, natural abundance 50.5%		
^{80m}Br	4.5 h	37 keV γ	Neutron source required
^{81}Br	Stable, natural abundance 49.5%		
^{82}Br	36 h	80 keV β^-, 0.78 MeV γ	Commercial suppliers or thermal neutron source
^{123}I	13 h	160 keV γ	Cyclotron produced
^{124}I	4 d	β^+, 0.6 MeV γ	Cyclotron produced
^{125}I	57 d	35 keV γ	Commercial suppliers
^{126}I	13 d	β^+, 0.87 MeV γ	Fast neutron source
^{127}I	Stable, natural abundance 100%		
^{128}I	25 min	2.1 MeV β^-, 0.45 MeV γ	Thermal neutron source
^{129}I	1.6×10^7 yr	50 keV β^-, 40 keV γ	
^{131}I	8 d	0.61 MeV β^-, 0.36 MeV γ	Commercial suppliers
^{132}I	2.3 h	1.53 MeV β^-, 0.67 MeV γ	Nuclide generator kits

of radiohalogens available commercially is rather more restricted, partly because of the short half-life of some of the isotopes. ^{77}Br and ^{123}I are both cyclotron-produced radionuclides, and therefore relatively expensive. ^{125}I is a very popular isotope because of its moderate half-life and low-energy γ emission (which allows shielding masses to be kept low). Most of the radiohalogens are supplied in the form of halide ions in aqueous solution or as the elemental halogen in CCl_4 solution.

There are two common approaches to the labelling of organic molecules with radiohalogens: halogen exchange labelling and small group substitution. Halogen exchange labelling is relatively straightforward for labelling with bromine or iodine, in that a (non-radioactive) halogenated compound is refluxed with a solution of the radiohalogen in an appropriate solvent until either sufficient exchange has taken place or equilibrium has been reached, e.g.:

$$CH_2ICOOH + {}^{125}I \xrightleftharpoons{\text{reflux}} CH_2\,{}^{125}ICOOH + I$$

A modification of the method is necessary for chlorine labelling (and may also be useful for brominations) and is achieved by using aluminium trichloride (or $AlBr_3$) as the radiohalogen source, e.g.:

$$C_2H_5Br + Al^{82}Br_3 \rightleftharpoons C_2H_5{}^{82}Br + AlBr^{82}Br_2$$

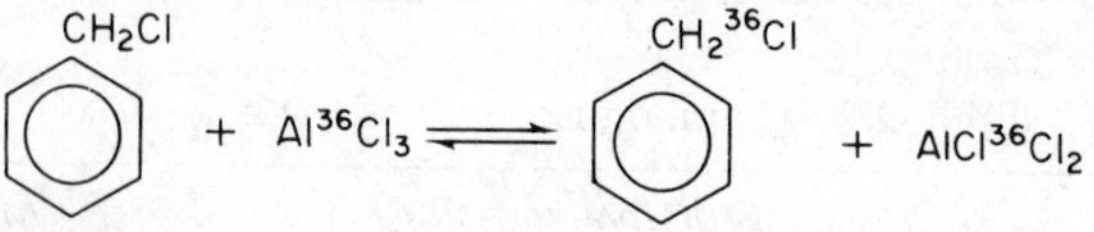

For many exchange reactions which use the AlX_3 reagent the exchange is fast enough at room temperature to make refluxing unnecessary.

The principal disadvantage of isotope exchange labelling is the relatively low specific activity achievable in practice. In some cases it is possible to overcome this limitation by exchanging one halogen for a radioisotope of a different halogen, so that the radiolabelled product may be isolated in carrier free form by some form of chromatographic separation, e.g.:

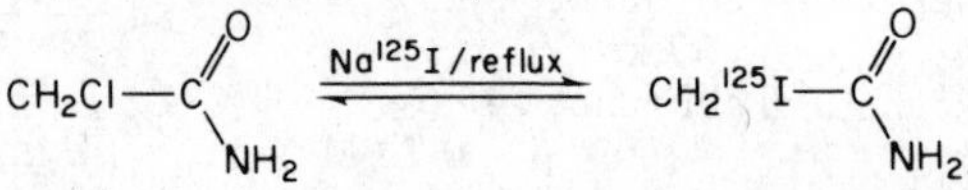

A wide range of examples of radiolabelling by halogen exchange is given by Murray and Williams, who demonstrate that a number of different reagents may be used in such reactions, including HX, LiX, NaX, AlX_3, SnX_4 and X_2.

Small group substitution reactions provide an alternative means of radiohalogenation, and the wide choice of reaction conditions has resulted in these being more useful than isotope exchange in practice. One commonly used technique is experimentally similar to the halogen exchange labelling method: a compound, which differs from the desired product only by the absence of a halogen atom, is refluxed with a solution of the radiohalogen until the reaction is complete.

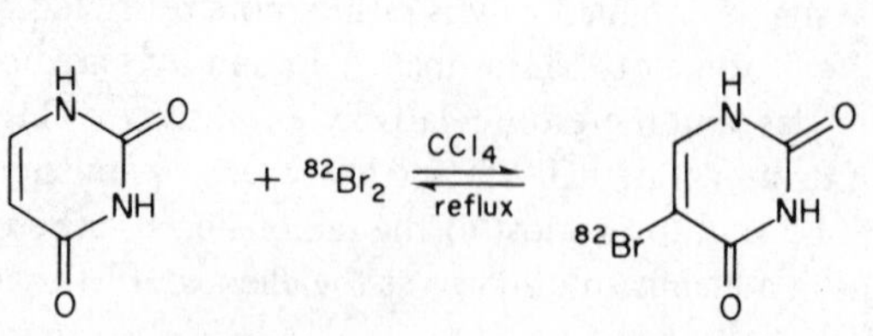

The preparation of labelled alkyl iodides may be achieved by a well known variation involving a small quantity of phosphorus as a catalyst, e.g.:

$$C_2H_5OH + {}^{125}I_2 \xrightarrow{\text{P reflux}} C_2H_5{}^{125}I$$

For the iodination of biomolecules in aqueous solution a range of oxidative labelling procedures has evolved. Some of these techniques were originally designed for protein iodination, although it has been found that many small aromatic molecules may be rapidly and conveniently labelled using similar procedures. Probably the simplest iodination reaction may be brought about by mixing buffered aqueous solutions of the target biomolecule and radioiodide ion in the presence of a chemical oxidising agent such as hydrogen peroxide or chloramine-T (a common name for $CH_3 \cdot C_6H_4 \cdot SO_2N(Na)Cl.3H_2O$). Iodination of proteins generally occurs at tyrosyl residue, i.e.:

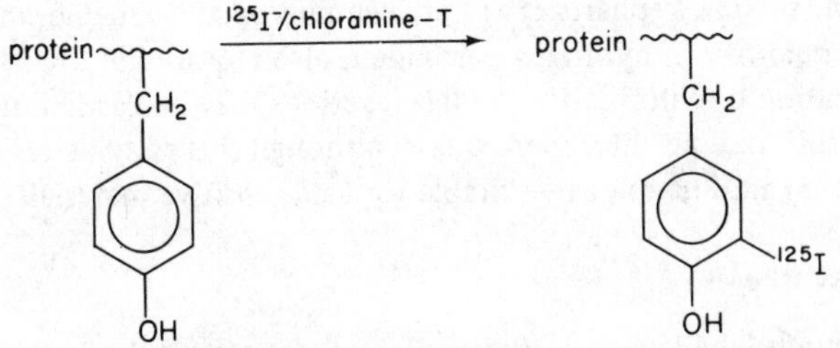

and not surprisingly the reaction works efficiently for tyrosine itself as well as for a wide range of other water-soluble aromatic and heterocyclic compounds, e.g.:

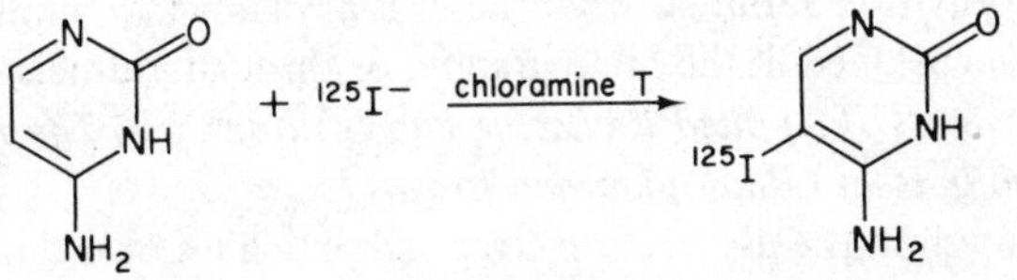

Oxidative bromination using chemical oxidising agents is also useful, although the range of small molecules which may be radiobrominated using chloramine-T is more limited. The real limitation of radiohalogenation of biomolecules using chemical oxidising agents is that the oxidising conditions may damage the precursor or the desired product, particularly when large proteins are involved. For example, chloramine-T and hydrogen peroxide have been reported to diminish the specificity of proteins labelled for use in radioimmunoassay. For this reason milder oxidative labelling methods have become increasingly popular for the radiohalogenation of biomolecules.

Electrolytic iodination has been used for protein labelling and is reported to have less effect on antibody specificity than the chemical iodination procedures. Essentially the method involved the electrolysis of a mixed solution of the target protein and radioiodide ions, the latter being oxidised at the anode and released for direct iodination, e.g.:

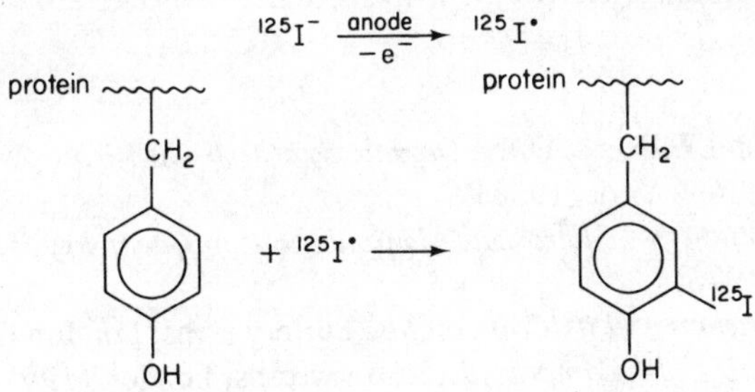

An alternative method, which has the advantage of not requiring special apparatus, is enzymatic labelling. Many proteins and a range of small aromatic or heterocyclic compounds may be iodinated by reaction with radioiodide ion in buffered aqueous solution (~ pH 7) under the catalytic influence of the enzyme lacto peroxidase. The enzyme may be used in solution or immobilised on a solid state support (such as Sepharose) or between microporous membranes; in either case a small quantity of hydrogen peroxide is also required in the solution. For radiobromination or chlorination similar reactions may be carried out in the presence of the enzyme chloroperoxidase, although this enzyme requires a rather low pH (~ 2–3) and may not be suitable for acid-sensitive materials.

9.5 Other syntheses

Most compounds labelled with other radionuclides are synthesised by more specialised routes than are necessary in the case of the univalent radiolabels or, in the particular case of ^{14}C, where an unusually large number of active precursors is available. Information about such syntheses is available in a number of the specialist texts mentioned in the bibliography, or through journals such as *The International Journal of Applied Radiation and Isotopes*, and *The Journal of Labelled Compounds and Radiopharmaceuticals.*

The rapid growth in the use of short-lived radionuclides over the past decade has required the development of an increasing number of very rapid labelling procedures to overcome the problem of the time required for more traditional syntheses. For example, novel biosynthetic methods using intact organisms and plant materials have been reported for the rapid preparation of ^{11}C-labelled sugars ($t_{1/2}$ for $^{11}C \sim 20$ minutes), and enzymatic methods for the production of several ^{13}N-labelled amino acids ($t_{1/2}$ for $^{13}N \sim 10$ minutes). Several fluorination reagents have been developed for rapid ^{18}F labelling ($t_{1/2}$ for $^{18}F \sim 2$ hours) including $Ag^{18}F$, $K^{18}F/CH_3COOH$ mixtures and $^{18}F^-$ held on ion-exchange resin, in each case the subsequent labelling reaction relies on a nucleophilic displacement mechanism.

Radiolabelling with short-lived radionuclides is a rapidly growing field of interest, partly because of the obvious difficulty in obtaining such materials from commercial suppliers. There will probably be major developments in this field in the coming decades as short-lived radioisotopes become more widely available via nuclide generators and (relatively) inexpensive neutron generators and cyclotrons.

Bibliography

Murray III, A. and Williams, D. L., *Organic Synthesis with Isotopes*, Vols I and II, Interscience, New York (1958)

Ma, T. S. and Horak, V., *Microscale Manipulations in Chemistry*, J. Wiley & Sons, New York (1976)

Evans, E. A., *Tritium and its Compounds*, Butterworths, London (1974)

Catch, J. R., *Carbon-14 Compounds*, Butterworths, London (1961)

Catch, J. R., *Labelling Patterns: their Determination and Significance*, Review 11, The Radiochemical Centre, Amersham (1971)
Wahl, A. C. and Bonner, N. A. (eds), *Radioactivity Applied to Chemistry*, J. Wiley & Sons, New York, 323-364 (1951)
Journal of Labelled Compounds and Radiopharmaceuticals, Published quarterly by J. Wiley & Sons, New York
International Journal of Applied Radiation and Isotopes, Published monthly by Pergamon, Oxford

10 Purity, Stability and Storage of Labelled Compounds

10.1 Purity and radiochemicals

Most scientists are familiar with the concept of 'purity' as it is applied to conventional chemical substances. Normally expressed in percentage terms, the chemical purity of a material may be defined as the percentage of the mass of material present which is in the specified chemical form. Thus a sample of toluene may be described as 98% CH_3-C_6H_5. Unfortunately when dealing with radiolabelled compounds the situation is not so straightforward. For example, impurities may or may not be radioactive, or may even be labelled with 'the wrong radioisotope'. Furthermore, the amount of impurity present in a given sample may vary with time or, particularly in the case of materials labelled with short-lived radionuclides, may become relatively more important as the principal radiolabel decays.

One of the major uses of radiolabelled molecules is as tracers for chemically similar non-radioactive compounds. When used in this way the fraction of molecules in a bulk of material which are labelled with a radionuclide may be very small. For example, a sample of mono-iodo-tyrosine labelled with ^{125}I to a specific activity of 100 MBq mol^{-1} has only one molecule in 10^{11} which actually contains the ^{125}I isotope. Where ratios of this magnitude are involved, purity clearly assumes a high level of importance not generally encountered with chemical materials.

For a close consideration of purity of radiolabelled compounds it is convenient to define two additional terms. The first is radiochemical purity, which is the percentage of total radioactivity present in a sample which is in the specified chemical form. For example, a sample of ^{14}C-labelled toluene may have a radiochemical purity stated to be 95% $^{14}CH_3$-C_6H_5, indicating that for each Bq of activity within the sample 0.95 Bq originate from $^{14}CH_3$-C_6H_5 molecules and 0.05 Bq from other sources. The specified chemical form referred to may encompass such factors as the position of the label or the enantiomorphic form of the radiolabelled compound.

The second term is radionuclidic purity, which is the percentage of the total radioactivity present as the specified radionuclide. For example, a sample of ^{131}I, in the form of, say, an aqueous iodide solution, may have a radionuclidic purity stated to be 90% ^{131}I, indicating that 90% of the activity arises from the decay of ^{131}I, while the other 10% arises from the decay of other radionuclides. It is important to remember that a stated radionuclidic purity implies nothing about the chemical form of radionuclides present in a sample.

Both the above definitions refer to relative radioactivities, and indeed it is the activity which is normally estimated in experiments involving labelled compounds. Neither definition provides any information about the quantity of radioactive material present in a sample compared with the inactive material. If this information is required then both the chemical purity and the specific activity of the sample must be known in addition to the radiochemical purity. For example, a sample of toluene of 99% chemical purity, labelled with ^{14}C to a specific activity of 10^9 Bq mol^{-1} and with a radiochemical purity given as 99% $^{14}CH_3\text{-}C_6H_5$, has only a small fraction of its molecules with the composition $^{14}CH_3\text{-}C_6H_5$. This fraction is given by

$$\frac{\text{chemical purity} \times \text{radiochemical purity} \times \text{specific activity of sample}}{\text{specific activity of pure } ^{14}\text{C}}$$

which in the above example is $\sim 5 \times 10^{-4}$.

10.2 Determination of radiochemical purity

In practice it is the radiochemical purity of a sample of radiolabelled material which is most likely to be the important factor in a radiochemical experiment. Major suppliers of radiolabelled compounds usually aim at a radiochemical purity of better than 95% for a freshly supplied product, and in fact most ^3H and ^{14}C-labelled compounds are supplied at better than 98%. Unfortunately radiolabelled materials have an annoying tendency to approach lower radiochemical purities as they are stored, particularly if the storage conditions are not carefully controlled. For this reason many users of radiochemicals prefer to determine the radiochemical purity of a sample immediately before use.

The determination of radiochemical purity is not normally difficult, although there is no shortage of pitfalls for the unwary. The radiochemical purity of most materials can be ascertained by some form of radiochromatography, in particular radiogas chromatography or radio HPLC. These techniques have the advantage that, given the availability of suitable detection systems, the chemical purity may be estimated at the same time. The older chromatographic methods of paper and thin layer chromatographies are not normally used for purity estimation for organic materials because of the risk of volatilisation of the sample material or the impurities, particularly since the latter may be present at very high specific activities.

For appropriately labelled compounds, reverse isotopic dilution analysis may be used with inactive carriers free from detectable impurities. In some instances reverse isotopic dilution analysis (RIDA) may be more useful than a chromatographic method for the detection of enantiomorphs as impurities, and RIDA may also be used for the assay of specific impurities in a labelled sample.

Unfortunately none of the above methods is suitable for determining the specificity of labelling in a compound which contains many atoms which are isotopes of the radionuclide. In this case the normal procedure involves some kind of chemical degradation of the sample to isolate the relevant portions of the molecule, followed by the determination of the activities of each portion. These procedures can be time consuming and difficult to carry out with precision, although fortunately such action is largely restricted to materials labelled with tritium or carbon-14, and in these cases the specifically labelled compounds available commercially tend to retain their specificity even though the overall radiochemical purity may decrease with time.

10.3 The stability of labelled compounds

Leaving aside the specific activity of a radiolabelled compound, which will clearly decrease as the radionuclide decays away, a factor of major importance in the use of radiochemicals is that the chemical, radiochemical and, in some cases, even the radionuclidic purity of an active compound may all decrease on storage. It is essential to understand some of the factors which produce these changes if storage conditions are to be chosen to minimise the effects.

The chemical and radiochemical purities of a labelled compound may decrease if the compound is thermodynamically unstable, or if the radioactive decay or accompanying emission induces decomposition. Problems of thermodynamic stability are not restricted to radiolabelled compounds, and the same criteria will apply to compounds whether they contain a radionuclide or not. For example, some materials require storage at low temperature or in the absence of oxygen or water vapour, and such requirements should not be overlooked simply because a compound is radiolabelled. Similarly normal chemical processes will occur within samples containing radiolabelled compounds. For example, an aqueous solution containing a ^{14}C-labelled carbonate or bicarbonate will exchange CO_2 with the atmosphere. While such exchange would normally be of no interest in inactive solutions, the result can be rather startling with radioactive material. In the case of a very high specific activity ^{14}C-carbonate solution, for example, a few days exposure to the atmosphere can result in almost total loss of activity from the solution.

By far the most important factor affecting the stability of radiolabelled compounds is the induction of decomposition by the radioactive decay. Decay may induce decomposition in surrounding molecules in two ways, normally termed primary decomposition and secondary decomposition. Primary decomposition is the decomposition of a molecule by direct interaction with radiation (α, β, γ or

X radiation). Such interaction will normally result in the production of an ion or radical which will undergo further reaction to end up as an impurity which may or may not contain a radionuclide.

Secondary decomposition is the decomposition of sample molecules by reaction with ions or radicals within the mass of the sample. The ions or radicals which cause secondary decomposition include both the daughter species formed in radioactive decay and the species produced by the interaction of emitted radiation with the mass of the sample. In practice it is secondary decomposition which makes the largest contribution to changes in purity of most radiolabelled compounds and which compels careful consideration of storage conditions if its effects are to be minimised.

The extent of both primary and secondary decomposition depend on the amount of energy deposited within the sample by radioactive decay. Nuclides which decay by α emission deposit large amounts of energy within a relatively short distance of the decay site, both because α decays are normally quite energetic processes involving energies of several MeV, and because α particles are densely ionising with very short mean ranges. γ-emitting nuclides deposit relatively little energy within most samples because the γ photons interact only rarely with the matter through which they pass, and in most cases γ photons are expected to escape from the sample almost entirely. However, this comment must be treated with caution because γ-emitting radiolabels are often used in situations where the labelled material has a very high specific activity (e.g. in nuclear medicine procedures) or in situations where the γ photon energy is quite low (e.g. ^{125}I) and has a relatively high probability of causing ionisation.

For β-decay isotopes the situation is rather more complex because of the different types of emission involved and the wide variation in the energies of emitted particles. β^- particles lose most of their energy (are most densely ionising) towards the end of their tracks, so that energetic β^- particles (> 1 MeV) may escape from a solution and enter the atmosphere or the container wall before depositing the bulk of their energy. On the other hand weak β^- emitters, such as ^3H, yield β^- particles of very short range (fractions of a millimetre), so that virtually all of the decay energy is deposited in the vicinity of the decaying nuclide.

10.4 Quantitative aspects of radiation-induced decomposition

The rate of decomposition of a radiolabelled compound is strongly dependent on the nature of the compound concerned and the physical state of the sample. The study of the chemical effects of the interaction of radiation with matter is called radiation chemistry from which we obtain a concept which can be helpful in discussing the sensitivity of different samples to decomposition induced by their own radioactive decay. This is the *G(–M)* value for a compound in a given state, and represents the number of molecules of the compound irreversibly altered per 100 eV of energy absorbed by the sample. The sample may be a chemically pure compound or a solution of the compound and, not surprisingly, a single compound may have several $G(-M)$ values depending on the state of the sample, the nature of any solvent and the concentration of the compound in a solvent. In practice most ^3H or ^{14}C labelled compounds stored under reasonable conditions (*see* below) have $G(-M)$ values between 0.1 and 10.0. However, there are several instances of materials yielding $G(-M)$ values up to 1000, so that it becomes difficult to offer reliable guidelines.

If a $G(-M)$ value is available for a particular labelled compound in the form of interest to the user then the percentage decomposition per day, P_d, for that compound may be evaluated using:

$$P_d = 1.43 \times 10^{-19}\, \bar{E}\, f\, s\, G(-M)\ \%\,d^{-1} \qquad (10.1)$$

where $\bar{E}$ is the mean energy of the emitted radiation in eV; f is the fraction of this energy deposited within the sample; and s is the specific activity of the labelled material in Bq mol^{-1}.

For the particular cases of tritium and carbon-14 labelled compounds, equation 10.1 reduces to

$$P_d = 7.82 \times 10^{-16} .s.\ G(-M)\ \%\,d^{-1} \text{ for } ^3H$$

and

$$P_d = 7.15 \times 10^{-15} .s.\ G(-M)\ \%\,d^{-1} \text{ for } ^{14}C$$

where f has been taken as unity in each case. A typical example of the use of such equations may be seen in the following: a solid sample of D-Glucose-6-^3H at a specific activity of 17.3×10^{12} Bq mol^{-1} has a $G(-M)$ value of 2.7 when stored at 0°C. Thus the sample undergoes decomposition at a rate of

$$P_d = 7.82 \times 10^{-16} \times 17.3 \times 10^{12} \times 2.7$$
$$= 3.65 \times 10^{-2}\ \%\,d^{-1}$$

or about 1% per month of storage.

$G(-M)$ values for a wide range of compounds stored under reasonable conditions (*see* below) were published in *Storage and Stability of Compounds Labelled with Radioisotopes*. (Unfortunately, at the time of writing, that publication is out of print, although presumably still available in many libraries). However, the dependence of $G(-M)$ values on storage conditions and specific activity has led most workers to determine their own $G(-M)$ values appropriate for individual compounds stored under particular conditions. Even so there are additional uncertainties associated with equation 10.1. For example, although it is normally assumed that $f = 1$ for low-energy β^- emitting isotopes, there is no generally accepted way of choosing an f value for a high-energy β^- decay or for γ-emitting radionuclides. Furthermore there are probably many factors which contribute to the decomposition in addition to the primary and secondary decomposition mechanism discussed above. Our present lack of understanding of such factors may result in wide discrepancies between predicted and actual decomposition of radiolabelled compounds.

10.5 Deterioration of radionuclidic purity

While the above discussion has been concerned with changes in chemical and radiochemical purity, there are some applications in which changes in radionuclide purity may be important. Two factors should be borne in mind when such variations are being considered. First, some radionuclides decay to daughter nuclides which are themselves radioactive. In some cases the daughter nuclide may be a different element from the parent (e.g. after α or β decay), in which case a re-purification immediately before use may restore the radionuclidic purity of the sample. However, in the case of metastable nuclides the daughter may be a radioisotope of the same element as the parent and may be difficult (or virtually impossible) to remove.

If there is a sufficiently large difference in half-life between the parent and daughter radionuclides, then the build up of the daughter may have little effect on the radionuclidic purity of a sample until the parent has been stored for many half-lives. In the case of ^{99m}Tc, which decays to ^{99}Tc, the half-lives are 6 hours and 2.1×10^5 years respectively. In a sample of ^{99m}Tc which has been stored for 6 hours (and assuming the original radionuclide purity to have been 100% ^{99m}Tc) the number of ^{99}Tc nuclei will be about the same as that of the remaining ^{99m}Tc nuclei. However, the radionuclide purity of the sample, as we have seen, is defined as

$$\frac{\text{activity due to } ^{99m}\text{Tc}}{\text{total activity of sample}} \times 100\%$$

which is given by

$$\frac{\text{No. of } ^{99m}\text{Tc nuclei} \times \lambda_{99m}}{\text{No. of } ^{99m}\text{Tc} \times \lambda_{99m} + \text{number of } ^{99}\text{Tc} \times \lambda_{99}} \times 100\%$$

or, ignoring the number of ^{99}Tc nuclei which have decayed in the six hour period,

$$\frac{\lambda_{99m} \times 10^2}{\lambda_{99m} \times \lambda_{99}} \equiv \frac{t_{1/2}\,(^{99m}\text{Tc})^{-1} \times 10^2}{t_{1/2}\,(^{99m}\text{Tc})^{-1} + t_{1/2}\,(^{99}\text{Tc})^{-1}}\ \%$$

So the radionuclide purity is

$$= \frac{6^{-1} \times 10^{-2}}{6^{-1} + 2.1 \times 10^5 \times 365 \times 24^{-1}}\ \% = 99.99999967\%\ ^{99m}\text{Tc}$$

If the difference in half-life between parent and daughter is small, then the change in radionuclidic purity resulting from decay may be much more important. For example, ^{69m}Zn ($t_{1/2}$ = 13.9 hours) decays to ^{69}Zn ($t_{1/2}$ = 55 minutes), which in turn undergoes β decay to stable ^{69}Ga. A sample of ^{69}Zn, originally 100% radionuclidically pure, will after, say, 12 hours have a radionuclidic purity given by

$$\frac{\text{activity of } ^{69m}\text{Zn}}{\text{total activity}} \times 100\%$$

$$= \frac{\text{No. of } ^{69m}\text{Zn nuclei} \times \lambda_{69m} \times 100\%}{\text{No. of } ^{69m}\text{Zn} \times \lambda_{69m} + \text{No. of } ^{69}\text{Zn} \times \lambda_{69}}$$

For storage times which are long compared with the half-life of the daughter it is reasonable to assume that a decay equilibrium has been established (see chapter 2), i.e.

$$\text{No. of } ^{69}\text{Zn nuclei} = \text{No. of } ^{69m}\text{Zn} \times \frac{\lambda_{69m}}{\lambda_{69}}$$

Consequently the radionuclidic purity of our sample simplifies to

$$\frac{\lambda_{69m}}{\lambda_{69m} + \lambda_{69m}} \times 100\% = 50\% \ ^{69m}\text{Zn}.$$

The second factor of importance in connection with changes in radionuclidic purity is the ratio of storage time to the half-life of the prime radionuclide. If a short-lived radionuclide contains a longer-lived impurity, then the radionuclidic purity of the short-lived material will decrease on storage, as the prime isotope decays away and the activity of the impurity becomes a larger part of the total. This is a problem which may be encountered with a range of radionuclides, although the best known examples are of short-lived, γ-emitting nuclides for which the choice of synthetic routes is limited. For example, ^{123}I ($t_{1/2} \sim 13$ hours) is a cyclotron produced radionuclide–prepared by α-particle bombardment of tellurium or antimony targets. During production small amounts of ^{124}I ($t_{1/2} \sim 4$ days) and ^{125}I ($t_{1/2} \sim 60$ days) are formed and may find their way into the ^{123}I sample. Freshly prepared samples may have a radionuclidic purity of $> 99\%$ ^{123}I, while the relative activity of the impurities will approximately double every 13 hours as the ^{123}I decays (the exact change will depend on the ^{124}I/^{125}I impurity ratio).

At first sight this example may seem of little importance. However, ^{123}I is an expensive isotope and that provides an incentive to use a sample for as long as the activity is adequate. Its decay, by photon emission, is unaccompanied by the emission of charged particles (which could cause radiation damage inside a human being) and this makes ^{123}I highly desirable for *in vivo* nuclear medicine procedures. The impurity material, ^{124}I, does emit charged particles (1.53 and 2.13 MeV β^- particles) and so can be a problem in ^{123}I samples which are growing old.

Fortunately most manufacturers of radionuclides choose production routes which result in isotopes of a high radionuclidic purity and which avoid the problem of a rapidly deteriorating radionuclidic purity.

10.6 The storage of labelled compounds

In the light of earlier comments on the radiation induced decomposition of radio-labelled compounds, it is clear that efforts must be made to minimise such decomposition. Of prime importance, especially for materials at high specific activities, is the correct choice of storage conditions. While advice on particular problems may often be obtained from commercial suppliers of radiochemicals, the following comments provide a guide to the factors which should be considered in deciding on storage conditions for radiolabelled samples.

Essentially there are three areas in which precautions may be taken to minimise induced decomposition. These are: (1) the dispersal of the active material; (2) the choice of storage temperature; and (3) the use of radical scavengers.

(1) The dispersal of active material

The extent of both primary and secondary decomposition may be reduced by dispersing the sample molecules, so that the chance of any one molecule encountering an emitted particle or a radical resulting from radiation damage is made as small as practical. For solid materials this generally means spreading the compound over a wide area, either as a powder or as a thin film on the inside surface of the container. For liquids, or solids which may be conveniently dissolved in a suitable solvent, a dilute solution can be made in a purified solvent. Non-aqueous solvents are preferred as these avoid the problems associated with the radiation induced decomposition products of water (such as OH. and H_2O_2). Benzene, methanol, and ethanol are the most commonly used dispersal solvents, although benzene has a freezing point which is rather high (5.5°C) compared with normal storage temperatures and care must be taken to ensure that the solvent does not crystallise out leaving a concentrated solution of labelled material which then suffers considerable decomposition.

Dispersal of active material within an inactive carrier of chemically identical material is also a means of reducing the rate of decomposition of the labelled compound. Of course, this is deliberately lowering the specific activity of the sample and this may be unacceptable for some applications. However, it is generally undesirable to store labelled compounds at higher specific activities than necessary.

(2) Storage temperature

As a general rule the rate of secondary decomposition may be reduced by storing samples at as low a temperature as possible. For most laboratories this means in a deep-freeze at ~ −20°C, and although in many cases a lower temperature would result in a lower decomposition rate, it is doubtful whether the provision of storage facilities at −78°C or −196°C would be economical.

There is, however, at least one important exception to the general rule. It has been observed that tritium-labelled compounds in aqueous solution actually undergo more decomposition when stored frozen than when stored at just above 0°C. Although the reason is not fully understood, it seems likely that this results in part from the clustering of solute molecules on freezing. The short range of the low-energy β^- particles emitted by ^3H nuclei gives rise to the concentration of radiation damage rather close to the decay site, and hence, on average, to a cluster of the sample molecules. Whatever the explanation the facts are clear; tritium labelled compounds in aqueous solutions are best stored at a low temperature, but above the freezing point.

(3) The use of scavengers

The rate of secondary decomposition of labelled compounds, particularly those in aqueous solutions, can often be reduced by the addition of a few per cent of a radical scavenger. Again the details of the mechanism of radical scavenging are not fully understood, but it seems clear that the scavenger species acts as a trap for radicals and fragments which may otherwise react with the labelled compound. Various substances have been used for this purpose, although the most general useful scavenger is probably ethanol, as this may be easily removed from the aqueous solution when required. Other scavengers include benzyl alcohol, sodium formate and cysteamine, although these are not as readily removed from solution as ethanol. It is not always essential to remove the scavenger from the solution before using the labelled compound, and in any case radiochemical stock solutions are often so greatly diluted before use that the presence of small amounts of scavenger may not interfere with the experiment.

Many commercially available ^{3}H and ^{14}C labelled compounds are stabilised by a few per cent of a scavenger–usually ethanol–and it is worth noting that if such materials are diluted for storage it may be necessary to add extra scavenger if the rate of decomposition is to be kept to a minimum.

(4) Other precautions

In addition to the above expedients for minimising the effects of radiation induced decomposition, there are a few other precautions which may be taken to reduce the extent of chemical decomposition during the storage of labelled compounds. For example, compounds which are sensitive to oxidation may well benefit from the removal of oxygen from the storage vessel or the solvent. Bearing in mind that the macroscopic quantity of high specific activity materials may be very small, the simple exclusion of oxygen–either by vacuum pumping or purging with nitrogen or argon–may be well worth while.

The use of antioxidants (such as hydroxy-butyl-toluene) may also be advantageous where these compounds do not adversely affect the chemical processes in the intended application. Stock solutions which could present a favourable habitat for the growth of micro-organisms (such as aqueous solutions of biomolecules) may also be protected by the addition of a bacteriostat, and 0.1% sodium azide solution is sometimes used for this purpose.

In general stock solutions of radiolabelled compounds are best stored in total darkness.

Bibliography

Bayly, R. J. and Evans, E. A. 'Stability and Storage of Compounds Labelled with Radioisotopes', *Journal of Labelled Compounds*, part 1 in **Vol. 2**, 1-34 (1966); part 2 in supplement to **Vol. 3**, 349-379 (1967)

Bayly, R. J. and Evans, E. A., *Storage and Stability of Compounds Labelled with Radioisotopes*, Review 7, The Radiochemical Centre, Amersham (1968)

Evans, E. A., *Self-decomposition of Radiochemicals*, Review 16, The Radiochemical Centre, Amersham (1976)

11 Health and Safety Aspects

In the early days of experimentation with radioactive materials the effects of radiation on man were unknown. High levels of activity were handled in conventional chemical laboratories, and the story is told of how Pierre Curie used to carry in his pocket an ampule containing such a high activity of radium that he could read by the glow of this now-notorious material. Tragically many of the early radiochemists were to find that the radiations emitted by the objects of their studies did have a profound effect on their bodies, causing burns, sickness, deformities and even death.

Since those days almost everybody has become aware of the potential dangers of radioactive materials. In this chapter it is hoped to introduce the reader to the basic safety measures which should be understood by anyone considering the use of radiochemical techniques in their work. The governments of most countries now exercise some degree of control over people using radioactive materials, largely with the intention of minimising the risks to those people and to the general public. In the UK control is exercised under the terms of The Radioactive Substances Act, 1960, and several orders and regulations which have followed it. Whilst it is not strictly necessary to study the Act itself, anyone wishing to work with radioactive materials in the UK should familiarise themselves with practical consequences of the legislation as outlined in the *Code of Practice for the Protection of Persons exposed to Ionising Radiations*. The comments made in this chapter are intended as an introduction to the health and safety aspects of radiochemical work, and do not represent a statement of any minimum legal requirement for those handling radioactive materials in any particular country.

11.1 Radiation dose

While radiochemicals do possess the usual properties associated with chemical compounds, e.g. some are poisonous, carcinogenic, inflammable, or explosive, the most important feature of radiolabelled compounds is that they emit radiations which may be harmful to man. In most radiochemical work the experimenter measures the activity of radioactive materials by counting the number of particles or photons emitted per unit time by the samples under investigation. However, in considering the effects of radiations on the human body, a more important quantity is the amount of energy absorbed by the body as the radiation interacts with it. The unit of dose is called the Gray and is defined as the amount of radiation which deposits 1 J of energy per kilogramme of material through which it passes. Thus 1 Gy of radiation deposits 1 J kg^{-1}. (An older unit is the rad; 1 rad = 0.01 Gy.)

An indication of the relation between radiation dose and radioactivity may be obtained by noting that a dose of approximately 10 mGy is received every hour by a person working 1 metre away from an unshielded ^{60}Co source of activity 10^9 Bq. (This would be an unacceptable level of radiation even for a few minutes. The source should be shielded by surrounding it with several centimetres thickness of lead).

It would be convenient if the effects of radiation on humans could be simply related to the radiation dose in Grays. Unfortunately this is not the case, largely because the biological effects of radiation arise not only from the amount of ionisation which the radiation produces within the irradiated subject, but also from the density of that ionisation. As we have seen α particles are more densely ionising than β^- particles, which are in turn more densely ionising than γ photons. Therefore the measure of radiation dose used in discussing the biological effects of radiation is rather different from the Gy, but has the advantage that the biological effects of different types of radiation may be related to a single measure of dose. The unit of equivalent dose used in discussing biological effects of radiation is the Sievert. An equivalent dose in Sv is defined as: dose in Gy × quality factor, where the quality factor is as given in *Table 11.1*.

Table 11.1 Quality factors of various types of radiation

Type of radiation	*Quality of factor*
X and γ rays, β^- particles with $E_{max} > 30$ keV	1.0
β^- particles with $E_{max} < 30$ keV (e.g. β^- radiation from ^3H)	1.7
α particles	10
Ions heavier than $^4He^{2+}$	20

Thus a dose (in Gy) of α particles is an order of magnitude more damaging to biological systems than a similar dose (in Gy) of energetic β^- particles or γ photons. This difference is taken into account by quoting the equivalent radiation doses in Sv. (An older unit is the rem; 1 rem = 0.01 Sv.)

The actual effects of a dose of radiation on a human depends on the dose (in Sv) and the time over which the dose was received. The effect of a dose of 5 Sv received over a short period of time (e.g. up to a few hours) is almost invariably death within a matter of days. The same dose spread over many years may not result in any obvious disturbances to the body (somatic effects), although it could still produce an effect on children born after the dose was received (genetic effects). Short duration doses of more than 0.5 Sv show a significant probability of inducing lukaemia or some other type of malignant disease in man. Note that there may be a time lag of many years between the receipt of a major radiation dose and the appearance of any evidence of harmful effects.

The biological effects of radiation are exceedingly complex and have been a subject of study for more than half a century. The International Commission on Radiological Protection (the ICRP) keeps under review the available data on the effects of ionising radiations. From time to time the ICRP issues recommendations on the maximum permissible doses (MPDs) which individuals working with radioisotopes, individual members of the public and indeed whole populations may receive without being at 'significant risk'. As more information has become available over the years these MPDs have been gradually reduced.

The governments of most countries have regulations limiting the maximum permissible doses which persons working with radioisotopes and others may receive. These MPDs may differ in detail from the ICRP recommendations, so that readers should obtain from the authority responsible for implementing the legislation which covers them, details of the MPDs which apply in their situation.

11.2 Dose monitoring

We all receive a radiation dose at a fairly steady rate from radioactive materials in the environment and from cosmic radiation which reaches the surface of the Earth. For example, on the streets of most cities in England or the northeastern USA a standard human receives a dose of slightly under 1 mSv per year. A chest X ray taken with good quality equipment will also deliver a dose of about 1 mSv, although in this case within a fraction of a second. It is generally believed that the effects of radiation on the body are cumulative. A man aged 30 who has had 10 chest X rays during his lifetime, will have received a dose of at least 40 mSv. By comparison the maximum permissible dose in the UK for a person occupationally exposed to radiation is 1 mSv per week, with an emergency situation once-in-a-lifetime exposure limit of 250 mSv.

The above doses are whole-body doses (based on the assumption that all parts of the body receive the same dose). Persons occupationally exposed to radiation may legally receive somewhat higher doses to their hands and forearms.

Persons working in radiochemical laboratories may be exposed to radiation hazards in two distinct forms: an external radiation dose, from emissions by radioactive materials in the laboratory; and an internal radiation dose, from radioactive materials which may have been ingested, inhaled or absorbed through the skin (particularly through damaged skin, cuts, etc.). Legal limitations on the MPDs which a designated 'radiation worker' (a person occupationally exposed to ionising radiation) may receive have resulted in most laboratories operating under a system whereby both doses received by individuals are routinely monitored.

Generally each laboratory is the responsibility of a radiological safety officer, who supervises radiation dose monitoring and arranges for radiation workers to have medical examinations in the event of any significant radiation exposure.

The external radiation dose may be monitored by equipping each worker with a dosemeter which is worn on the person and which records the radiation dose received—by the dosemeter. Such dosemeters are normally worn on the chest or abdomen. One such device is the film badge dosemeter, which is a small strip of photographic film wrapped in a light-proof cover and held in a pin-on or clip-on holder. Exposure to radiation is detected when the film is processed, and may be quantified by the degree of blackening of the processed emulsion.

In recent years there has been a tendency to replace film badge dosemeters by thermo-luminescent dosemeter badges (known as TLD badges). These consist of holders containing small sheets of a material which, after exposure to ionising radiation, give off visible light photons when heated. With appropriate control of the heating conditions, the number of photons emitted is proportional to the total radiation dose received by the material. Several companies now market complete systems for the preparation of TLD badges and for their processing after use to obtain the dose in Gy received by each badge. Finger monitors, which consists of film or TLD badges in convenient finger holders, are also useful for monitoring the radiation dose received by fingers and hands—areas of the body which may be subject to additional exposures when handling labelled materials in simple chemical operations.

As an alternative to the film or TLD badge, a number of electronic pocket dosemeters are available commercially. Most of these are based on simple G–M detectors and record the total amount of ionisation induced within the detector. Some are able to display the rate at which a radiation dose is being received. All three types of dosemeter are sensitive to X and γ radiation, and the film and TLD badges will also respond to high-energy β^- particles. α and low-energy β^- particles are not detected by these types of dose monitor, as these radiations are unable to penetrate the wrappings or casings. However, this is not a serious problem in the sense that these latter radiations are also unable to penetrate clothing or skin, so that individual dosemeters do give a useful guide to the external radiation dose received by the body of the wearer. For the same reason it follows that there is little point in operating a system of individual dosemeters of the types described in laboratories where only low activities of ^{3}H and ^{14}C are used.

Electronic dosemeters are particularly valuable for use on occasions when a higher-than-normal risk of exposure is anticipated, as they can give an immediate indication of dose. Some relatively inexpensive pocket radiation monitors are available which, although not providing a quantitative dose reading, do give an audible warning on receipt of a high-radiation dose rate. Such instruments are invaluable in situations where a rapid response to a radiation hazard is desirable, e.g. where high activity γ-emitting materials have been delivered and require unpacking for dispensing.

Monitoring the internal radiation dose received by radiation workers is remarkably difficult. Some radioactive materials which gain entry to the body will localise in particular organs and remain there for years; for example, ^{90}Sr becomes incorporated in bone, where it remains until it decays. Other materials may become generally distributed and then excreted relatively quickly; e.g. ^3H_2O becomes absorbed in body tissue, but the amount remaining in the body is halved every 12 days—its biological half-life is 12 days. Little is known about the areas of localisation or the metabolism of the vast majority of radiolabelled compounds which could find their way into the body, so the guideline is always to ensure that any intake of radioactive material is kept to an absolute minimum. For example, mouth operated pipettes are never used in radiochemical laboratories, and eating, drinking or smoking is forbidden.

In the particular case of a suspected intake of a γ-emitting isotope of iodine, monitoring any emission from the thyroid gland with a γ-scintillation detector (e.g. a contamination monitor, *see* below) may be helpful. But generally where an intake of radioactive material is suspected, monitoring the urine for activity is the only practical way of testing for an abnormal internal radiation dose. In some laboratories urine samples may be collected routinely for monitoring.

Apart from efforts to monitor the doses received by radiation workers, it is normal practice for occasional medical checks to be made. In the UK the most common check of this kind is the blood test, in which the condition of a few millilitres of blood is analysed to ensure that no abnormalities have arisen. Blood tests are also carried out after any accidental exposure to significant levels of radiation.

11.3 Radiation and contamination monitoring

The monitoring techniques outlined above are normally the responsibility of a radiological safety officer whose main concern is the protection of people in and around a laboratory from any unnecessary doses of radiation. Not surprisingly the radiological safety officer will have an interest in the location of all radioactive materials in the laboratory, and in minimising the amount of radiation in any area used by people. There are two aspects of this work which are of considerable interest to the experimenter, not only because of possible health risks, but also because unnecessary radiation in a laboratory can interfere with the counting of samples. These two aspects are the shielding of radioactive materials to prevent significant radiation doses being received by people in the laboratory, and the location and removal of contamination of the laboratory and apparatus by unwanted radioactive material.

Radiations emitted by inadequately shielded radioactive materials or areas of radioactive contamination can be detected by the use of portable monitors. These are relatively inexpensive instruments which consist of a radiation detector and a small box of electronics containing an appropriate high voltage supply, amplifier and count rate meter. Radiation monitors are generally single units (i.e. the detector is built into the box of electronics) and most are based on G–M detectors. As a result they are only sensitive to X and γ radiation between energies of

~ 50 keV and ~ 2.5 MeV, although some instruments have a thin window which allows the detection of lower energy photons and high energy β^- particles. The monitor is held or placed in the region of the suspected radiation and indicates the count rate, or radiation level in mGy h^{-1}, on the rate meter.

Contamination monitors usually have the radiation detector in a separate probe unit, so that the detecting element can be moved into awkward places while the rate meter is watched. A variety of probes are used for detecting the different radiations. For example, simple, thin window G-M tubes are used for monitoring energetic β^- emitters, and NaI (Tl) crystal scintillation probes for monitoring γ emitters with γ photons above ~ 50 keV. For monitoring α or weak β^- emitters (e.g. ^{14}C or ^{35}S) special G-M probes with very thin mica windows are often used, although the detection efficiencies are rather poor. Slightly better results may be obtained using scintillation probes which have a phosphor covered with a very thin light-proof foil. The increased use of low energy photon emitters, such as ^{125}I, has led to the use of X-ray monitors for detecting emissions below the threshold of the more conventional NaI (Tl) scintillation probes. Recently some companies have offered thin window, thin crystal (~ 1 mm thick) NaI (Tl) scintillation probes for use in the 5–50 keV region, and these work very well for monitoring ^{125}I and other low energy photon emitters.

Most contamination monitors have simple rate meters reading in counts per second. In monitoring for contamination one generally wishes to estimate the activity of a spot of contamination, or the activity per unit area of an extended area of contamination. Relating the recorded count rate to an activity requires both a knowledge of the nature of the contaminant radionuclide (so that it is desirable to segregate working areas in which different radionuclides are used) and a regular calibration of the monitor against standards of known activity. For α and weak β^- emitters it is particularly important that the window of the probe is positioned as close as possible to the contaminated surface (preferably without actually transferring contamination to the probe), as the range of these particles in air may be only a few millimetres.

In addition to the above contamination monitors for the location of contamination in the laboratory, monitors are also used for checking for contamination on the hands or clothing of laboratory personnel. These are usually fixed monitors, located near the washing facility for people leaving the laboratory. The hand monitors normally give a straightforward 'safe' or 'wash' indication after the hands have been monitored, and the personnel monitor gives an alarm indication of any contamination detected as the probe is passed over the body.

Checking for surface contamination by tritium using the portable contamination monitors described above is an almost impossible task, unless the level of contamination is so high that a reasonable Bremsstrahlung count can be detected with an X-ray probe. The most commonly adopted solution to this problem is the wipe test, which is also useful for checking general low-level contamination of floors, benches, etc. A small filter paper is moistened with a solvent, rubbed over the suspected area in a single stroke, and placed in a vial of liquid scintillator cocktail for counting. From the detected activity and the area covered by the collecting wipe, some indication of the level of contamination may be obtained. Note that, in the absence of evidence to the contrary, it should be assumed that no more than one tenth of the removable contamination is detected by a wipe test, so that contamination levels should be multiplied accordingly.

At least one commercial instrument (the Spasciani CCSI surface contamination sampler, available from E.S.I. Nuclear) has refined the wipe test procedure so that it may be carried out reproducibly. A small vacuum cleaner draws air and dust into a filter paper pressed lightly on the suspected surface. The filter paper may be moved over the surface, sweeping out a precise area, and then be removed for counting.

Finally there is always the worry about airborne contamination in any radiochemical laboratory, particularly by HTO which is often used in relatively large activities for tritium labelling reactions. One company (Nuclear Enterprises) markets a portable radioactive gas monitor (RGMI) which measures the activity concentrations of α- or β^--emitting radioactive gases in air by drawing air through an ionisation chamber. An alarm circuit is provided to give audible or visual alarm indication at an adjustable level of contamination.

11.4 Classification of radionuclides and laboratories

While there are several classification systems for radionuclides, which differ because of the different aims of the systems, one system which should be familiar to all who work in radiochemical laboratories is the classification according to

Table 11.2 Classification of some radionuclides according to radiotoxicity per unit activity

Class 1	*High toxicity*
	^{239}Pu, ^{240}Pu, ^{241}Am, ^{210}Pb, ^{226}Ra, ^{233}U, ^{252}Cf
Class 2	*Medium toxicity*–upper subgroup A
	^{22}Na, ^{36}Cl, ^{45}Ca, ^{60}Co, ^{68}Ge, ^{90}Sr, ^{110m}Ag, ^{125}I, ^{131}I, ^{137}Cs, ^{224}Ra, ^{236}U
Class 3	*Medium toxicity*–lower subgroup B
	^{14}C, ^{18}F, ^{24}Na, ^{32}P, ^{35}S, ^{38}Cl, ^{42}K, ^{47}Ca, ^{47}Sc, ^{51}Cr, ^{52}Fe, ^{55}Fe, ^{59}Fe, ^{57}Co, ^{58}Co, ^{63}Ni, ^{64}Cu, ^{65}Zn, ^{67}Ga, ^{74}As, ^{76}As, ^{75}Se, ^{77}Br, ^{86}Rb, ^{85}Sr, ^{90}Y, ^{99}Mo, ^{99}Tc, ^{113}Sn, ^{132}Te, ^{123}I, ^{132}I, ^{131}Cs, ^{197}Hg, ^{203}Hg, ^{206}Bi, ^{239}Np
Class 4	*Low toxicity*
	^{3}H, ^{11}C, ^{13}N, ^{15}O, ^{68}Ga, ^{85}Kr, ^{87m}Sr, ^{99m}Tc, ^{113m}In, ^{129}I, ^{232}Th, ^{235}U, ^{238}U

radiotoxicity per unit activity. The complete list is published in the ICRP Publication 5 and, in the UK, in the Code of Practice. Only a representative selection of the radionuclides most commonly encountered is given in *Table 11.2*. It is clear that the heavy metal α emitters represent a class which must be considered to be highly radiotoxic relatively low activities can produce serious biological effects within man.

In the UK, laboratories are classified into three grades (A, B and C), and the grade of laboratory required for handling radionuclides is related to the amount of activity which may be present and the radiotoxicity class of the nuclides. The approximate requirements are shown in *Table 11.3* for laboratories in which normal chemical operations are to be performed. Actual requirements may be slightly different for two reasons: at the time of writing, regulations on activity limits are still quoted in curies; and stricter requirements are imposed for laboratories in which complex wet operations (with a high spill risk), or dry and dusty operations are performed.

Table 11.3 Activity ranges appropriate for use in laboratories of different grades

Radiotoxicity classification	*Laboratory grade* C	*Activity range* (Bq) B	A
Class 1	$< 10^5$	10^5–10^7	$> 10^7$
Class 2	$< 10^7$	10^7–10^9	$> 10^9$
Class 3	$< 10^9$	10^9–10^{11}	$> 10^{11}$
Class 4	$< 10^{11}$	10^{11}–10^{13}	$> 10^{13}$

Figure 11.1 The radiation trefoil. The internationally accepted symbol to indicate the actual or potential presence of ionising radiation.

Most modern chemical laboratories with non-absorbent working surfaces and at least one good induced-draught fume cupboard may be readily modified to grade C standard. Grade B laboratories are purpose designed to allow control of contamination, and require approved fume cupboards and laboratory air changing plant. Design recommendations for grade B laboratories have been published in ICRP Publication 5. Grade A laboratories are very sophisticated establishments in which radioactive materials are generally handled by remote control.

All radiochemical laboratories must have clearly marked controlled areas, i.e. those areas where radioactive materials are handled and in which a radiation hazard may be present. The internationally accepted symbol used to indicate the actual or potential presence of ionising radiations is the trefoil or 'tree', shown in *Figure 11.1* (although normally seen on a yellow background). This symbol is still occasionally encountered with its trunk uppermost—in which case it is upside down.

11.5 Decontamination

Large amounts of activity left on or in laboratory apparatus or on bench tops and floors represent a health hazard and are indicative of sloppy laboratory practice. But even relatively small activities in the wrong place can cause problems in the radiochemical laboratory. A few thousand Bq of activity spilled in a counter can render the counter virtually useless until the activity is removed. A similar amount in an item of glassware could lead to totally misleading results from experiments performed in that apparatus. The elimination of such contaminating activities is called decontamination and is an essential aspect of radiochemical work.

A particle of a solid ^{14}C-labelled material the size of a grain of salt may easily have an activity of more than 10^4 Bq, and the same activity of an ^{125}I-labelled compound may be invisible to the naked eye. All radiochemical work should be carried out over spill trays, so that the small amounts of activity which inevitably become spilled or dropped are at least confined to an area which can be relatively quickly monitored using a suitable contamination monitor. However, it is also desirable that areas around a site of radiochemical operations are regularly monitored. When areas of contamination are discovered, these should be marked and decontaminated as soon as convenient.

Decontamination of working surfaces, walls and floors, can usually be achieved by washing with a decontaminating detergent, although for polished surfaces it may be necessary to remove the polish with an organic solvent. Where contamination is caused by a radioisotope of a metal, the addition of a chelating agent (such as 0.5% EDTA or ammonium citrate) may improve the decontaminating efficiency of detergent solutions. If these techniques are inadequate it may be necessary to apply a mild abrasive cleaner to scratch off the contaminated surface, taking care to avoid the release of contaminated particles into the air.

Laboratory apparatus is not more difficult to decontaminate, but it can be more difficult to check for contamination where this is in the form of α- or low-energy β^--emitting radionuclides. This problem often arises with liquid scintillation counting bottles which are to be reused (i.e. the more expensive quartz or low ^{40}K glass variety). Several efficient decontaminating detergents are available commercially (e.g. Decon 90 and Lipsol Liquid) as are some indicating liquids specifically designed for checking for contamination in glassware. All-glass apparatus can also be subjected to the standard 'chromic acid' treatment, or washing with organic solvents, depending on the nature of the suspected contamination. Plastic laboratory ware may usually be cleaned by washing in dilute nitric acid.

Although rubber gloves should always be worn when handling radiochemicals, any activity which does get onto the hands can normally be removed by washing with soap and water or, if that fails, with detergent solution. A soft scrubbing brush is invaluable for fingernails etc., but should not be used on the skin with such vigour that the skin becomes abraded. Contamination on other parts of the body, should not be doused with water (unless this is essential because of corrosive chemical splashing) as such action may spread the contamination and make it more difficult to remove than the original localised activity. Decontamination of the person is a rather specialised operation and should be performed under the supervision of the radiolological safety officer. Radiochemical laboratories should have clearly defined procedures for action in the event of a human contamination problem—and indeed for any accident involving active material—and everyone working in the laboratories should find out what these procedures are before a problem arises.

Decontamination procedures should be undertaken whenever it is found that contamination of a surface has reached the derived working limit, which is a level of activity per unit area based on the hazard that a particular contaminant

Table 11.4 Derived working limits for surface contamination

Surface	*Derived working limit* (Bq cm^{-2})		
	Class 1 α emitters	*Other α emitters*	*Other radionuclides* (except ^3H)
Surfaces of human body	0.37	0.37	3.7
Surfaces within controlled areas	3.7	37	37
Surfaces not within controlled areas	0.37	3.7	3.7
Interior surfaces of glove boxes and fume cupboards in controlled areas		The minimum that is reasonably practical	

represents. The derived working limits for contamination by radionuclides of the various radiotoxicity classes are given in *Table 11.4*. (The multiples of 3.7 appear because published limits are still quoted in curies [1 Ci = 3.7×10^{10} Bq]). Tritium is not included because of the practical difficulty of detecting surface contamination by this isotope.

11.6 Disposal of radioactive waste

Most governments require establishments using radioactive materials to be licensed or registered. The licence or registration documents will normally contain some statement as to the quantity and nature of activity which may be disposed of as radioactive waste, and the procedures which may be used for such disposal. In addition a separate authorisation is required in some countries (as in the UK) for the disposal of radioactive waste.

It is important to realise that all discharges of radioactive material are subject to some form of control. For example, the discharge of radioactive gases or vapours from laboratory fume cupboards or the normal laboratory ventilation system is generally restricted. There are also limits on the amounts of activity which may be discarded as solid refuse or poured into the foul water drainage system.

Radioactive wastes may be classed as short-lived or long-lived, the distinction varying from one country to another, in some cases being based on the practicality of storing short-lived waste until its activity has been reduced to a threshold level. In the UK the discharge of any radioactive gases or vapours is discouraged, and where significant activities of gaseous materials are to be used, efforts may have to be made to scrub or filter exhausts, or to trap condensible vapours for chemical conversion to less volatile forms. Where the discharge of gaseous activity is permitted the radiation and contamination levels near the point of discharge should be monitored periodically.

Short-lived solid or liquid waste may be safely stored until its activity is low enough for authorised disposal. For longer-lived wastes the trend seems to be to encourage disposal as soon as possible (to minimise radiation and spill risks which would be associated with storage). Some disposal of solid waste by incineration or via the refuse system (for adequately packaged materials) may be permitted, as may some disposal of aqueous liquid waste via the foul water drainage system. In some cases, organic liquid wastes may be burned. The activity levels and classification (often even the identity) of the radionuclides which may be treated in these ways are specified by the terms of the waste disposal authorisation.

Other wastes must be disposed of by official radioactive waste disposal organisations, such as the UK waste disposal service which operates from Harwell. The disposal of radioactive waste can be an expensive operation.

When planning experiments involving radiolabelled compounds thought should be given to the disposal of radioactive waste. In the UK, it is an offence to accumulate or dispose of radioactive waste if authorisation has not been obtained from the Department of the Environment, or, if authorisation has been given, to dispose of waste in a manner which is not in accordance with that authorisation. Careful records must be kept of the purchase, dispensing and disposal of all radioactive materials, and these records must be available for inspection by government inspectors who have the authority to revoke licences and authorisations of establishments in which inadequate care is exercised in the management of radionuclides.

At the time of writing the UK government is considering changes in the law relating to radioactive materials. It seems likely that a number of the quantities discussed above may be modified during 1980, and the reader is advised to check on the regulations which apply to his particular establishment before commencing work with any radioactive materials.

Bibliography

On radiological safety

International Atomic Energy Agency publications
(available in the UK from HMSO, London)

A Basic Toxicity Classification of Radionuclides
(Technical Reports Series No. 15, 1963)

Basic Safety Standards for Radiation Protection
(Safety Series No. 9, 1967)

Manual on Safety Aspects of the Design and Equipment of Hot Laboratories
(Safety Series No. 30, 1969)

Monitoring of Radioactive Contamination on Surfaces
(Technical Reports Series No. 120, 1970)

Personnel Dosimetry Systems for External Radiation Exposure
(Technical Reports Series No. 109, 1970)

Safe Use of Radioactive Tracers in Industrial Processes
(Safety Series No. 40, 1974)

The Management of Radioactive Wastes produced by Radioisotope Users
(Safety Series No. 12, 1965) *and Technical Addendum* (Safety Series No. 19, 1966)

Safe Handling of Radionuclides, 1973 edn
(Safety Series No. 1, 1973)

International Commission on Radiological Protection publications
(Pergamon, Oxford):

ICRP Publication 2–*On Permissible Dose for Internal Radiation*
ICRP Publication 3–*On Protection against X rays up to 3 MeV, and β and α rays*
ICRP Publication 4–*On Protection Against Electromagnetic Radiation above 3 MeV*
ICRP Publication 5–*On the Handling and Disposal of Radioactive Materials*
ICRP Publication 9–*Recommendations of the ICRP*
ICRP Publication 12–*General Principles of Monitoring for Radiation Protection of Workers*
ICRP Publication 15–*Protection against Ionising Radiations from External Sources*
ICRP Publication 26–*Recommendations of the ICRP. A Summary*

UK Acts and Official Publications
(HMSO, London):

Radioactive Substances Act, 1960

The Ionising Radiations (Sealed Sources) Regulations, 1969
(S.I. 1969 No. 808)

The Ionising Radiations (Unsealed Radioactive Substances) Regulations, 1968
(S.I. 1968, No. 780)

Code of Practice for the Protection of Persons exposed to Ionising Radiations
in Research and Teaching (1968)

The Control of Radioactive Wastes, Cmnd 884 (1962)

Other publications

Radiological Protection in Universities,
The Association of Commonwealth Universities, London (1966)

Street, H. and Frame, F. R., *Law Relating to Nuclear Energy*,
Butterworths, London (1966)

Appendix: Names and Addresses of Miscellaneous Suppliers

Suppliers of radionuclides and radiochemicals

The Radiochemical Centre,
Amersham,
Buckinghamshire, UK

Isotope Production Unit,
AERE Harwell,
Didcot,
Oxfordshire, UK

New England Nuclear,
11 Chaucer Close,
South Wanston,
Winchester, Hants, UK

Suppliers of simple counting systems

E.S.I. Nuclear,
Klempfern House,
Holmesdale Road,
Reigate, UK

Nuclear Enterprises Ltd,
Sighthill,
Edinburgh, UK

Panax Nucleonics,
Trowers Way,
Holmethorpe Industrial Estate,
Redhill, Surrey, UK

Suppliers of γ-spectrometry equipment

Ortec Ltd,
Luton,
Bedfordshire, UK

Nuclear Data Inc,
Rose Industrial Estate,
Cores End Road,
Bourne End,
Buckinghamshire, UK

Canberra Instruments Ltd,
223 Kings Road,
Reading,
Berkshire, UK

Suppliers of liquid scintillation counters

Beckman-RIIC Ltd,
Eastfield Industrial Estate,
Glenrothes,
Fife, UK

LKB Instruments Ltd,
232 Addington Road,
Selsdon, South Croydon,
Surrey, UK

Kontron,
P.O. Box 188,
Watford, UK

G. D. Searle & Co. Ltd,
Nuclear Chicago Division,
Lane End Road,
High Wycombe,
Buckinghamshire, UK

Nuclear Enterprises Ltd,
Sighthill,
Edinburgh, UK

Suppliers of liquid scintillation cocktails, etc.

Koch-Light Laboratories Ltd,
Colnbrook,
Buckinghamshire, UK

LKB Instruments Ltd,
232 Addington Road,
Selsdon, South Croydon,
Surrey, UK

Fison Scientific Apparatus,
Loughborough,
Leicestershire, UK

Suppliers of equipment related to radiochromatography

E.S.I. Nuclear,
Klempfern House,
Holmesdale Road,
Reigate, UK

ICN Instruments,
277 Antwerpse Steenweg,
2800 Mechelen,
Belgium

Panax Nucleonics,
Trowers Way,
Holmethorpe Industrial Estate,
Redhill, Surrey, UK

Suppliers of safety products

Film badge service

National Radiological Protection Board,
Personal Monitoring Service,
Harwell, Didcot,
Oxfordshire, UK

TLD systems

Pitman Instruments,
Jessamy Road,
Weybridge, Surrey, UK

Radiation and contamination monitors

E.S.I. Nuclear,
Klempfern House,
Holmesdale Road,
Reigate, UK

Nuclear Enterprises Ltd,
Sighthill,
Edinburgh, UK

Mini-Instruments Ltd,
8 Station Industrial Estate,
Burnham-on-Crouch,
Essex, UK

Decontamination materials

Decon Laboratories Ltd,
Conway Street,
Hove,
East Sussex, UK

Index of Radionuclides

Subject Index